Beyond Numbers:
A Practical Guide to Teaching Math Biblically

God bless you!
Katherine K. Loop
Psalm 105:4

Katherine A. Loop
Christian Perspective ❖ Fairfax, VA

Copyright © 2005, 2007 by Katherine A. Loop and its licensors.
All rights reserved.

No part of this guide may be reproduced in any manner or by any means without the written permission of the author. Brief quotations in articles and reviews are permitted.

First Printing: 2005
Second Printing: 2007

Published by
Christian Perspective
Fairfax, VA
www.christianperspective.net

ISBN 10: 0-9773611-1-X
ISBN 13: 978-0-9773611-1-3

Graphic Designer: Adrienne Butler
Illustration Artist: Jeff Cortazo

Unless otherwise marked, "Scripture taken from the HOLY BIBLE, NEW INTERNATIONAL VERSION (North American Edition). Copyright © 1973, 1978, 1984 by International Bible Society. Used by permission of Zondervan Publishing House."

When marked NASB, "Scripture taken from the NEW AMERICAN STANDARD BIBLE ®, Copyright © 1960, 1962, 1963,1968, 1971,1972, 1973, 1975, 1977, 1995 by The Lockman Foundation. Used by permission."

*Dedicated to my Aunt Marie
And to the many other
Homeschool moms who will read this book.
May you and your children
Be blessed by seeing God in math.*

Contents

Preface	vii
Acknowledgements	ix
Chapter 1: Where Did Math Come From and Why Does It Work?	11
Chapter 2: Math Points Us to God	17
God's Greatness	17
God's Wisdom and Care	18
Chapter 3: Math's Practicality	21
Math Helps Us Explore Creation	21
Math Leads to Useful Inventions	23
Math Assists Us in Our Day-to-Day Life	25
Chapter 4: Math Is Not Neutral!	29
Chapter 5: Harm to the Heart	35
Chapter 6: Adopting a New Heart toward Math	39
Math Methods from the Past	43
Arithmetic Then…	43
Geometry Then…	44
The Joy of Biblical Math	45
Chapter 7: Preparing to Teach Math Biblically	51
Remind Yourself of God's Truths	51
Examine Your Goals	52

Chapter 8: Teaching Math Biblically 55
  General Guidelines 55
    View Math Biblically 57
    Use Math Practically 59
    Putting It All Together 61

Chapter 9: Ready, Set, Now What Do I Do? 65
  In Your Curriculum 66
  In Resources 66
  In Your Own Life 67

Chapter 10: Curriculums and Resources 69
  Guidelines to Follow 69
  Curriculums 71
  Biblical Reference Books 76
  Supplemental Math Books 77

Chapter 11: Overcoming the Difficulties 79

Parting Thought 83

Appendix A: Idea Notebook 85
  Common Ways We Use Math 85
  Common Projects That Use Multiple Math Concepts 90
  A Few Basic Ways Math Is Used in Science 90

Appendix B: Bibliography 91
  Modern Textbooks 91
  Older Textbooks 93
  Supplemental Math Books 95
  Miscellaneous 96

Preface

Math was always a strong subject for me. My father and grandmother are CPAs. My mom, who homeschooled me since first grade, majored in math. Because my entire family loved math, I learned to enjoy it as well. I progressed rapidly, completing Algebra 2 by ninth grade.

Yet, while I did well in math, I did not view math biblically. I recognized God's presence in other subjects, but not in math. Science clearly pointed to the Creator. History vividly demonstrated God's sovereignty over the affairs of man. But math? I did not understand how God had anything to do with math.

As I progressed to the higher levels of math, I began to wonder why I needed to learn some of these math concepts. How would I ever use geometry or advanced algebra in my own life? I learned the advanced math concepts like one would learn a fascinating game, but I saw few practical applications for what I learned.

In my senior year, my view of math changed entirely after I read a book titled *Mathematics: Is God Silent?* by James Nickel. I started the book rather skeptically, but soon could not put it down. All my old ideas about math crumbled. Math was *not* neutral. God *was* in math. I did not have to simply believe what my textbooks told me. Math, even geometry and algebra, had meaning and application in real life.

After I graduated, God prompted me to write this simple, nontechnical guide to biblical math. I have spent the last two years researching and writing this guide. My appreciation for math and my understanding of how it points to God has continued to grow as I have explored the Scriptures, numerous math textbooks, and *Truth in the Transcendent*, a book on the Bible and math by Larry Zimmerman. Giving piano instruction has also taught me a great deal about teaching and contributed to many of the ideas I will share in this guide. Many of the other teaching ideas came from watching my mom learn and grow as she taught my brother and me.

In this guide you will discover what the Bible says about math. You will explore where math originated and why it works. You will see the harm in the way most textbooks teach math. You will walk through all aspects of teaching math biblically, from preparing and choosing a curriculum, to overcoming obstacles. Throughout the process, you will learn to look beyond numbers and to behold God in math.

May the Lord use this guide to encourage and equip you to look to Him in everything, *including* math.

KATHERINE LOOP

Acknowledgements

I am amazed and humbled by the number of people who have helped make this guide possible. I would like to a take moment to thank them all for their assistance and to let them know what a blessing and answer to prayer they have been to me.

First, I wish to thank James Nickel for his book *Mathematics: Is God Silent?* This book showed me the error in "neutral" math and taught me to approach math as God's handiwork. For this as well as for his encouragement with this project I am truly grateful.

Second, I wish to thank my family who went out of their way to help me. I especially wish to thank my mother, who has tirelessly helped in so many ways, and my grandfather Charles Loop, who has devoted hours to this project and whose honest opinions have helped me focus and simplify my thoughts.

I owe a special thanks to three talented people, without whom this book would still be an idea: Liz Harp, the editor, Jeff Cortazzo, the illustrator, and Adrienne Butler, the graphic designer. Words do not adequately express my thanks to you for freely giving so many hours of your time and God-given talents.

It would be impossible for me to mention by name all those who have, in one way or another, helped me with this guide. Many moms gave of their time to review this project prior to its publication, while others graciously made their math textbooks available for review. The librarians at the National Education Library granted me access to and assisted me in viewing the library's rare 1800 math textbook collection. Numerous people all over the country upheld this project in their prayers and encouraged me to keep going. Many others have indirectly influenced this work by impacting my spiritual life over the years. Thank you to everyone for all you have done!

Above all, I want to thank the Lord for opening my eyes to His presence in math. Compiling this guide has been an exciting journey into God's faithfulness. Without Him, neither math nor this little guide would be possible.

Chapter 1:
Where Did Math Come From and Why Does It Work?

$7^1/_8 + {}^3/_8 = 7^1/_2$

$88 \div 2 = 44$

$4 \cdot 12 = 48$ $\qquad y = 2x^2 + 8$

$a + b = c$

$8 < 9$

$^1/_3 \neq {}^7/_8$

  We are all familiar with math. As toddlers we learn to count. In first grade, we learn to add and subtract. We eventually discover multiplication, division, fractions, and decimals. As teenagers, we study algebra, geometry, trigonometry, and perhaps calculus. Some of us even go on to study math in college and graduate school. As adults, all of us use math in various activities and ways.

  And whenever we think of or use math, we also think of God and how math honors Him. Right? Not really. Most of us have been taught to look at math as a list of numbers and rules. We picture math as a set of number problems like the ones at the beginning of this chapter. Although we may see God's hand in history and science, we do not even look for God in math. We do not see any eternal harm or good in math.

  We view math as a neutral subject. Neutral means indifferent or, "not engaged on either side; not aligned with a political or ideological grouping."[1] We regard math as a subject "not

---
[1] *Webster's New Collegiate Dictionary*, 1974 ed., s.v. "Neutral."

engaged" and "not aligned" with either biblical or worldly thinking. Believing that math is independent from God, we approach math as a "safe" subject—a subject we can all see the same way, regardless of our religious beliefs. After all, the equation one plus one equals two ("1 + 1 = 2") works the same way for a Christian, Muslim, Buddhist, Hindu, or Atheist. Math is a subject of numbers and facts, and most of us think facts are neutral.

But is math really indifferent and neutral? Is such a thing even possible? The Bible warns us that we are in spiritual warfare (Ephesians 6:10-18). It urges us to guard our heart and to test the spirit behind what we are taught (1 John 4:1, Proverbs 4:23). The Bible does not mention any neutral ground. According to the Bible, nothing can really be neutral. Everything will be presented in either a biblical or a worldly fashion.

So how can math be presented biblically? What does God have to do with math? Join me now in looking at an equation we are all familiar with: "1 + 1 = 2." We will examine *where* this equation came from and *why* it works. As we examine this equation's origins and ability to work, we will discover some startling truths that apply to every area of math.

Mathematicians throughout history have developed various theories to explain the origin and consistency of addition. Some have speculated that addition exists by sheer chance. Others have claimed that man created addition and that addition works because man designed it to work.

Most modern textbooks do not even attempt to offer an explanation for addition's existence. Throughout my schooling, not one of my textbooks ever explained how addition originated or why it works. I eventually came to the conclusion that addition, along with all other math facts, is an eternal, self-existent truth.

The Bible gives us a radically different explanation for addition. While we will not find the word "addition" in the Bible, the Bible offers us principles that apply to addition, as well as every other aspect of math. Look at what just two Bible verses reveal.

The verse, "For by him *all* things were created" (Colossians 1:16), tells us where addition originated. It tells us that God created *all* things. The word *all* includes everything, even math. This does not mean that God created the symbols 1 and 2. Man developed those symbols. But those symbols represent a real-life principle called addition that is embedded in everything around us—a principle God created.

Let me briefly illustrate. We have often taken an object, added another object to it, and were then left with two objects (like when we took one cookie and one piece of pie and ended up eating two desserts). It does not matter what we attempt to add, objects always add in the same way. An accountant adds dollars, a baker adds doughnuts, a retailer adds inventory, and an electrician adds wattage. Yet addition works the same way for all of these people. Why? Because all things add in a regular, precise manner. We refer to this regularity as the addition principle.

God created this principle that prevails throughout creation. When God created all things, He determined how they would operate. He chose to make everything—from dollars to watts—operate according to the same reliable principle we call addition.

Man has developed different ways to express on paper this consistency God placed around us. Throughout history, cultures have used different symbols to represent quantities. For instance, the Romans used Roman numerals (I, V, X, etc.) instead of our current Arabic system (1, 5, 10, etc.). However, man has never *created* anything in math. He has merely *developed* different ways of representing the orderly way things add. Addition originated with God.

Now that we know *where* addition originated, let us look at *why* addition works. We will find the answer in the very next verse of Colossians, which reads, "In him *all* things hold together" (Colossians 1:17). Since the word *all* includes addition, God not only created addition but also continually holds addition together.

How does God hold addition together? We saw above that addition was a name for the predictable way objects combine. God is the One who keeps objects combining consistently. Thus, He is the One holding addition together.

Can you imagine what would happen if objects did not always combine in the same way? If the equation "$1 + 1 = 2$" only worked some of the time? Math as we know it would be impossible! It would be pointless to memorize addition facts and solve problems on paper if those facts and problems did not consistently work in the physical world. Ponder for a moment what a miracle it is that math *always* works. Whether we are on the moon or in our own backyard, one plus one *always* equals two. Math is consistent because God consistently holds every part of the universe in its place. God's faithfulness in holding this universe together ensures us that objects will *always* add the same way and that the equation "1 + 1 = 2" will always work.

The Bible reveals a lot about addition. We have only looked at two verses and already we know:

- Where addition came from (God created it); and

- Why addition works (because God faithfully sustains the universe).

The principles we have discovered about the addition equation "1 + 1 = 2" apply to the rest of math. Like addition, all of math is a way of recording and expressing the laws and relationships God created. Math works because God faithfully holds everything in place. Math is not independent from God. It is not neutral. Math's very existence and ability to work is dependent on God's faithfulness in holding everything together!

God's faithfulness not only makes math useful to us, but it also communicates an important message. The Bible tells us that God chose to establish a covenant with the "fixed laws" of heaven and earth (which would include mathematical laws like addition). His faithfulness in keeping His covenant with these laws demonstrates the same faithfulness He will show in His covenant of redemption.

> If I have not established my covenant with day and night and the fixed laws of heaven and earth, then I will reject the descendants of Jacob and David my servant.
>
> (Jeremiah 33:25b, 26)

The fact that God holds everything together according to fixed laws demonstrates that He is a covenant-keeping God. His consistency in keeping one plus one equaling two serves as a continual testimony to us of His loyalty, faithfulness, and loving kindness.

Just think, every time you use math to count or add something, you witness the fact that God is still consistently sustaining the universe. It is as though math were shouting out to you, "God is still in charge! He is faithful. He will keep His covenant of redemption with you."

Math is really a testimony to God's faithfulness and power. Math should continually remind us of God's consistency and trustworthiness. After all, one plus one *always* equals two because an awesome, never-changing God created and sustains all things.

◆◆◆◆◆◆◆

Math is not a neutral subject. As we understand biblically where math came from and why it works so consistently, we find that God is not only in math, but He created and sustains math. Instead of setting God aside when we approach math, let us set aside our old ideas about math and approach math expecting it to glorify the Lord.

Chapter 2:
Math Points Us to God

Since God created and sustains math, we should expect to see Him in math. Much like a painting reveals many of its painter's characteristics, math reveals many of God's characteristics. As we saw in the last chapter, math's consistency testifies to God's faithfulness. Here are a few other ways math points to God.

God's Greatness

Throughout math, we see an incredible system at work. Just think how long it takes us to learn even basic math skills. Most of us spend years trying to grasp a small percentage of math's many rules.

As mentioned in the last chapter, math records the orderly way that God determined things would interact throughout the universe. God created the complex structure of mathematics and holds it in place. Think about what this means for a moment. God created a universe that interrelates in such a complex and precise manner that mathematicians spend a lifetime trying to understand and record it. God's greatness knows no bounds!

Our finite minds can hardly comprehend some portions of math, such as infinite numbers. Yet God holds infinity in the palm of His hands. The Bible tells us that His thoughts toward us are infinite in number (Psalm 139:17, 18; 40:5). Infinite numbers can only exist because a God of infinite proportions created the universe. As hard as it is for us to comprehend, God is more infinite than infinity![1]

---

[1] God created the infinite galaxies of space. Even the highest heavens cannot contain Him! "But will God really dwell on earth? The heavens, even the highest heaven, cannot contain you. How much less this temple I have built!" (1 Kings 8:27).

Math's complexities should cause us to turn our eyes to God in awe and wonder, for God created and holds math in place. Math serves as a reminder to us of how much greater God is than we could ever imagine. He is truly beyond our understanding (Job 36:26).

God's Wisdom and Care

Math not only shows us God's greatness; it also testifies to God's wisdom and care. Math displays aspects of the universe that our limited senses cannot readily observe. As we use math in conjunction with science to explore the universe, math unveils a wondrously detailed design. This design reminds us that our great God is also a wise and caring God who cares for the tiniest details of His creation.

For example, math exposes something about a honeycomb that the naked eye misses. We can see that the bee stores honey in honeycombs made out of hundreds of tiny hexagons. This fact does not appear noteworthy until we bring math into the equation. If we analyze various shapes with differential calculus, we will discover that, of all the shapes in the universe, the hexagon holds the maximum content in the minimum space. God designed the bee to store its honey in the most efficient manner possible.[2] What a caring, thoughtful Creator we have!

Math uncovers some amazing design within our own bodies. A computer once did the math necessary to determine what shape would allow a red blood cell to best diffuse oxygen throughout the blood stream. Surprise! You have thousands of red blood cells flowing through your veins right now in the absolutely ideal shape.[3] By revealing this hidden order, math reminds us that God, not random chance, created us.

---

[2] James D. Nickel, *Mathematics: Is God Silent?* 2001 rev. ed. (Vallecito, CA: Ross House Books, 2001), 211.
[3] Allen L. Gillen, *Body by Design* (Green Forest, AZ: Master Books, 2001), 76.

Math Points Us to God

Math often exposes order our ears cannot even hear. By using math to analyze sound, scientists discovered that every time we hear a sound, we actually hear an orderly yet invisible wave. Even more incredibly, these waves all express a pattern orderly enough to be recorded algebraically.[4] Who would have guessed that God hid such fascinating order in sound?

Even examining a common sunflower mathematically reveals some surprising news. Did you know that the seeds in a sunflower are arranged in two patterns? Regardless of how many seeds the sunflower contains, the number of seeds will be distributed between the two patterns in the same mathematical proportion. This precise proportion enables any given sunflower to hold the maximum number of seeds.[5]

These are just a *few* examples of how math makes the complex design God placed around us visible. When we use math to explore the world, we will never cease to be amazed at the remarkable order it uncovers. Over and over, we will find ourselves awed at our Creator's wisdom and care. As we see His care over creation, it should encourage our own hearts that we can trust Him, for "If that is how God clothes the grass of the field, which is here today and tomorrow is thrown into the fire, will he not much more clothe you, O you of little faith?" (Matthew 6:30).

◆◆◆◆◆◆◆

Whether we are studying math or using math to investigate creation, math continually urges us to trust God. Within math, we see a complex system that testifies to God's greatness. When using math to explore the physical world, we discover an intricate design that points us to God's wisdom and care.

Every part of math causes us to exclaim with the Psalmist, "Great is the LORD and most worthy of praise; his greatness no one can fathom" (Psalm 145:3).

---

[4] Nickel, *Mathematics: Is God Silent?* 238-239.
[5] *This proportion has been named the Fibonacci sequence and is also present in many other flowers and parts of nature.*

Chapter 3:
Math's Practicality

Biblical principles affect more than just the way we see math. They also affect the way we use math. Once we realize that God created and sustains math, we will no longer regard math as meaningless symbols on a piece of paper. We will recognize that those symbols represent principles in the physical world and will consequently no longer be content to use math merely in a textbook. We will want to begin to use math practically in our own lives.

By creating a methodical universe and giving us the mental capacity to record that universe mathematically, God gave us a valuable provision to help us in our lives here on earth. We can think of math as a God-given, multi-purpose tool. Much like a shovel helps us dig deep into the ground and plant gardens or build buildings, math helps us explore the universe to new depths and develop useful inventions. Much like an ever-handy screwdriver or pocketknife assists us in thousands of little activities, math assists us in many daily tasks. In this chapter, we will take a look at a few of the practical ways math helps us.

Math Helps Us Explore Creation

Math helps us explore creation. It shows us the design hidden from our senses, such as the design in beehives, blood cells, sound waves, and sunflowers. Math also aids in understanding many other aspects of creation.

For example, have you ever wondered how we know about atomic structure? No one has ever seen atoms. They are too small. We know everything we know about them from—you guessed it—math! Scientists used mathematical principles coupled with experimentation to discover the structure of atoms (nucleus, protons, and electrons) and to understand the principles by which atoms operate.

Or have you ever wondered how we know the distance to a star? Obviously, we cannot simply measure that distance with any physical tool of measurement. We use math to estimate a star's distance.

Throughout history, men have used math to describe the consistent way the universe operates. Sir Isaac Newton used math to record the steady attraction objects have toward earth (called the law of gravity). He then applied his law of gravity to find the weight of the sun.[1] Johannes Kepler used math to discover the way planets orbit before modern spacecraft made it possible to actually see the planets in orbit firsthand. The ancient Mayan civilization used math to keep track of the movement of the heavenly bodies.[2] From the beginning of time, math has aided man in investigating the earth.

In many different ways, math helps us better comprehend the extraordinary world God created. In the process, it points us to the greatness of the God who created both the microscopic order of the atom and the vast distances of space.

---

[1] Irving Adler, *The Giant Golden Book of Mathematics: Exploring the World of Numbers and Space* (NY: Golden Press, 1960), 86.

[2] The Mayans kept detailed almanacs chronicling astronomical movements and basing calendars off these movements. Michael Hoskin, ed., *Cambridge Illustrated History: Astronomy* (Cambridge, UK: Cambridge University Press, 1997), 14-17.

Math Leads to Useful Inventions

Nearly all inventions rely on math and the laws (i.e., consistent ways God holds things together) math uncovers. Math is instrumental in designing bridges. Computers are based on a mathematical system of zeros and ones. The strings in a piano and other stringed instruments all mathematically relate to one another. The list of mathematically-based inventions goes on and on.

Often, as we use math to help us explore creation, math indirectly leads to inventions. For example, understanding the atom enables us to predict chemical reactions, thus opening the door for developing all types of inventions. Knowledge of the atom has aided in manufacturing synthetic materials such as plastics, explosives such as the atomic bomb, various medical procedures such as x-rays, and even household items such as super glue.[3] Since electricity is based on atoms and how they interact, knowledge of the atom has given us a better understanding of electricity, leading to many of the electrical gadgets we have today. By teaching us about the atom, math has affected every field and aspect of modern life in one way or another.

Likewise, the discovery of the Law of Gravity has served a critical role in numerous inventions, especially in the aviation field. Because we can determine the pull of gravity, designers are better able to design aircraft, spaceships, and rockets to cope with the force of gravity they will encounter.

The use of mathematics in practical inventions is not limited to our Modern Era.[4] Many ancient civilizations, including the Egyptians and Greeks, used various types of sundials (which are based on math) to keep track of time.[5] The Egyptian pyramids demonstrate incredible mathematical usage, as do many of the old European cathedrals and buildings. If you have any doubt that math was used practically in the past, just con-

---
[3] *Super glue is based on molecular bonding. See Neil Schlager, ed., How Products Are Made: An Illustrated Guide to Product Manufacturing, vol. 1 (Detroit: Gale Research Inc., 1994), s.v. "Super Glue."*
[4] *For a further listing of different ancient inventions, see Peter James and Nick Thorpe, Ancient Inventions (New York: Ballantine Books, 1994).*
[5] *The New Encyclopedia Britannica, 15th ed., s.v. "Sundial." and R. Newton Mayall and Margaret W. Mayall Sundials: Their Construction and Use, 3rd ed. (Cambridge, MA: Sky Publishing Corp., 1994).*

sider the following list of inventions the Greek mathematicians in Alexandria, Egypt produced over two thousand years ago:

> Pumps to bring up water from wells and cisterns, pulleys, wedges, tackles, systems of gears, and a mileage-measuring device no different from what may be found in the modern automobile were used commonly. Steam power was employed to drive a vehicle along the city streets in the annual religious parade. Water or air heated by fire in secret vessels of temple altars was used to make statues move. The awe-struck audience observed gods who raised their hands to bless the worshippers, gods shedding tears, and statues pouring out libations. Water power operated a musical organ and made figures on a fountain move automatically while compressed air was used to operate a gun. New mechanical instruments, including an improved sundial, were invented to refine astronomical measurements.[6]

Quite an impressive list. Pumps, gears, mileage-measuring devices, steam powered vehicles, moving statues, musical organs, a compressed air gun—the Alexandrians found numerous practical uses for math!

NOTE

Alexandrian Greek mathematics differed drastically from the Classical Greek mathematics we will discuss later. While the Classical Greek mathematicians rarely used math practically, the Alexandrians, as you can see, used math quite extensively. Unfortunately, though they used math practically, the Alexandrians enthroned man's reason and missed out on seeing God in math. The Alexandrians serve as a warning to us not just to teach math as a practical tool. The practical element alone does not make math biblical.

---

[6] Morris Kline, *Mathematical Thought from Ancient to Modern Times* (New York: Oxford University Press, 1972), 103, quoted in Nickel, *Mathematics: Is God Silent?* 39.

Math Assists Us in Our Day-to-Day Life

In addition to providing us with useful inventions, math assists us in many different situations we encounter throughout our lives. Whatever our occupation, we will find math a useful tool.

Archimedes, a famous Alexandrian Greek mathematician, frequently used math to solve real-life dilemmas. Once, a king wanted to know whether some silver had been added to a gold crown he had ordered. Silver weighs less than gold, so Archimedes knew that if the crown had silver in it, it would be larger (and take up more space) than a solid gold bar of the same weight. By comparing the space the crown took in a bucket of water with the space taken by a solid gold bar of equal weight, Archimedes proved that the crown contained silver. According to legend, Archimedes also used mathematically-designed mirrors to reflect and intensify the sun's rays onto enemy warships, causing the ships to catch fire.[7]

Although most of us will not end up solving a golden mystery or setting fire to a fleet of ships, we can still use math to solve problems in our own life. For instance, we can measure the doorway to see if the new couch will fit through it *before* the deliveryman arrives. Below are some other common ways you may have used math to help you with everyday dilemmas.

While shopping

Rounding can be helpful when trying to purchase items within a specific budget. My mother, especially when paying for items with cash, always kept a running estimate in her head to avoid last minute surprises. She unconsciously rounded the price of each item she placed in her cart and kept a mental tally of the total.

Shopping uses much more math than just rounding. I remember being awestruck the first time I shopped for groceries by myself when I discovered how many different brands of noodles the store carried. Which should I buy? I grabbed the cheapest package I could find. My mom

---
[7] Nickel, Mathematics: Is God Silent? 40.

later pointed out to me that the least expensive is not necessarily the best deal. Had I used math to find the price per ounce, I would have realized that I paid more per ounce for my noodles than I would have had I chosen a different package.

Keep your eyes open next time you go shopping and see how many different math concepts you use. You may be surprised!

While cooking

Now you have finished your shopping. You come home to a hungry family. You throw together a batch of cookies and  decide to double the recipe so you can bring some cookies to soccer practice. You mentally double $3/4$ and know you need $1^1/_2$ cups of flour. Your $1/2$ and $3/4$ measuring cups are still being washed from lunch, so you use your 1 and your $1/4$ measuring cups instead, remembering to fill the $1/4$ cup twice. You have just added and multiplied fractions many times.

While repairing or improving your home

Home repairs and improvements require numerous math concepts. Just the other day my mother mixed wallpaper remover with water. If she had not mixed the solution in the correct mathematical proportions, the solution might have been too strong and damaged the wall.

If you have ever replaced a floor, you have used a lot of math. Finding the area of the floor, figuring out how much material to order, comparing prices between stores and different types of flooring, calculating the cost of grout and other supplies, and factoring in the cost of labor all require basic math.

Many other home improvement projects also require math. Determining how many rolls of wallpaper to purchase, figuring out what size piece of wood will cover the attic access, calculating how many square/linear feet of countertops to replace, or computing how large a vanity will fit in the bathroom are just a few other home improvements that involve math.

When buying a car or house

Math definitely comes in handy when buying a car or house. Math can help you figure out what your monthly loan payment will be. It can also aid in comparing interest rates and financing options.

We all use math frequently in our daily lives. Math is a useful tool for you and your children, not just for engineers or rocket scientists.

♦♦♦♦♦♦♦

Seeing the many ways math can be used to help us explore the earth, invent, and live day-to-day is exciting. Math is *not* a meaningless pile of numbers. It is a real, useful tool!

Yet math would not be useful at all if it did not always work, would it? We could not use math to invent or help us in real life if math only worked some of the time. Math is useful because it *always* works. And math *always* works because God consistently holds all things together. Math's usefulness points us to God's faithfulness!

Chapter 4:
Math Is Not Neutral!

When I first began to see that math testified to God, I began to wonder why I had not seen God in math before. Why had I gone through school viewing math as nothing more than a list of rules and numbers? How could I have been so blind? Looking back, I see that I was oblivious to God's presence in math because I viewed math as a neutral set of facts. I simply believed whatever my math books taught me. In my mind, math was a neutral subject, a subject that contained neither biblical nor worldly thinking.

I did not realize that, by viewing math as a neutral subject (a subject "not engaged" and "not aligned" with either biblical or worldly thinking), I was really embracing worldly thinking. I was learning math as a subject independent from God. If math is independent from God, then all the incredible order and consistency in math points us, not to the Creator, but to math and human intellect. This view of math is not just unbiblical; it is actually naturalistic and humanistic. It teaches us that we can trust the creation (math and human intellect) instead of the Creator. My "neutral" view of math was really a very worldly view, steeped in unbiblical thinking.

When you think about it, we cannot really view anything neutrally. The way we see everything will be automatically sifted through a biblical or a worldly philosophy. Even facts are interpreted through a belief system.

For example, consider the media. Newspapers and newscasts purport to report facts. Yet a conservative and a liberal reporter present facts differently. Even when reporting on the same subject, they will often choose to include different facts in their reports. They will unconsciously slant their presentation to their belief about the subject. If you doubt this, just compare a conservative and liberal paper around election time. Somehow, each paper will manage to portray the facts in a way that favors the candidate it prefers. Philosophies, or beliefs, come through in the way reporters select and report facts.

For another example, suppose an evolutionist and creationist were both to present the fact that they had found a fossil of a previously unknown animal. They would both report this fact in a radically different way. One would describe the fossil as the missing link, while the other would describe it as an animal that has gone extinct since the flood. One presentation presents the fossil from the worldly philosophy of evolution; the other presents it from the biblical philosophy of a world-wide flood.

This same thing happens in math but in a more disguised way. Whether we choose to acknowledge it or not, math facts are presented based on an overall belief system. This belief system will either be biblical or unbiblical. There is no other option.

Everything we study either presents God's view or the world's view. We could summarize these two views by the words "dependent" and "independent." God's Word teaches us that we are dependent on Him while all other philosophies claim that, to one extent or another, we can operate or please God on our own, independently from God.

As Christians, we are in continual warfare. We constantly have to fight our tendency to live and think independently from God. The Bible never excludes math from this combat zone. In math, as in everything else, we need to guard against independent thinking—any sort of thinking that encourages us to trust ourselves, operate independent of God, or view something with our own human reason.

Consider the following math lessons.[1] As you read them, ask yourself whether they present a biblical, dependent view of math and life or a worldly, independent one.

> Counting is a math skill that we learn early. Counting by ones, we say the numbers 1, 2, 3, 4, 5, 6, . . .
>
> These numbers are called **counting numbers**.[2]
>
> Searching for a missing factor is called **division**. A division problem is like a miniature multiplication table. The product is inside the box. The two factors are outside the box. One factor is in front and the other is on top.[3]

The facts stated in the above math lessons are all undoubtedly true. Counting is a math skill we learn early. One, two, and three are called counting numbers. Searching for a missing factor is called division. These lessons certainly appear spiritually neutral.

But look beyond the numbers and facts presented and ask which view of math—a dependent or an independent one—is being taught. While the above lessons present true math facts, they also leave out critical information. We are not told *why* these facts work nor are we taught how to use them in real life.

---

[1] Disclaimer: the following example lessons as well as the ones that appear later in the book were chosen from curriculums I had on hand and that I felt accurately represented typical math presentations. It is not my intent to condemn these curriculums over others. It is my intent to show the reader how "neutral" math lessons really present math through a worldly philosophy. This is true for nearly every math curriculum, not just the ones from which the examples were taken.
[2] Stephen Hake and John Saxon, Math 65: An Incremental Development, 2nd ed. (Norman, OK: Saxon Publishers, 1995), 1.
[3] Ibid., 74.

These lessons actually teach the student to view math as an independent, self-existent truth. Rather than presenting counting and division as principles created by God, they leave the student thinking of counting and division as man-made rules. The student does not see counting and division as practical, God-given tools. Instead, he is left viewing counting and division as neutral facts—facts independent from God.

The student reading the above lessons is being taught to think about math as something independent from God. This view of math goes directly against Scripture and subtly presents a completely unbiblical philosophy on life. Think about it. If math is independent from God, then it does not require God's help to work. Yet math works with miraculous consistency. We can clearly rely on math. If God is not responsible for math's ability to work, who is? By simply presenting facts without providing an explanation for their existence, this lesson leaves us thinking that either math works all by itself, or else that mathematicians somehow invented math and enable it to work.

Either way, something or someone other than God has received the credit for math's consistency. Math has been used to point us, not to God's faithfulness, but to math's faithfulness and man's ability to reason independently from God.

Do you see the subtle but dangerous philosophy hidden in these math facts? Teaching math as something independent from God really stems from the worldly philosophy that claims we do not need God—that we can determine truth and explain the universe on our own, using human reason and mathematical principles.

Let us look at one more example of a typical math lesson. Once again we will find that, by ignoring God, this lesson subtly indoctrinates the student in an independent view of math and life in general.

> A property describes the way something is. We can't change properties. We are stuck with properties because they are what they are.[4]

---
[4] John H. Saxon, Jr., *Algebra 2: An Incremental Development*, 2nd ed. (Norman, OK: Saxon Publishers, 1997), 8.

Math Is Not Neutral!   33

Again, there are some true facts in this math lesson. We cannot change properties. Yet, why can we not change properties? Why do properties work?

"We are stuck" and "they are what they are" clearly give the impression that math properties sustain themselves. They leave us viewing a property as something independent from anything or anyone else.

In contrast, the Bible teaches us that a property is really just one of the "fixed laws" by which God has chosen to sustain the universe. Far from being independent *from* God, a property is entirely dependent *on* God! When we study properties, we ought to be praising the Lord for His faithfulness. Like the other lessons, this apparently "neutral" lesson robbed God of His proper place in math.

Do you see how the above math lessons were not really spiritually neutral? Each one, by presenting math as something independent from God, subtly taught a worldly philosophy on life. They taught us to view math through *independent* eyes. These same independent eyes applied to other areas of life teach us that we can trust ourselves, we can determine truth apart from God, and we can save ourselves or please God on our own. The independent thinking in these math lessons, though subtle, is very dangerous.

In contrast, the Bible teaches us that math is dependent on God—that God created and sustains it. In fact, the Bible's whole message is one of dependence. Over and over, the Bible urges us to stop trusting our works, our wisdom, and ourselves. It continually stresses our need to depend on Christ alone in every area of our life. The Bible even compares us to a branch and tells us that, much like a branch can do nothing without the vine, we can do nothing apart from God.[5] When taught biblically, math should echo this message of dependence and encourage us that we can depend on our faithful Lord.

❖❖❖❖❖❖❖

---

[5] John 15:4-5

Math cannot really be presented neutrally. It will be presented in one of two ways:

- *Dependent* on God, i.e., something He created and sustains; or
- *Independent* from God.

Math either points toward God or away from Him. It either encourages us to depend on God or to live independently from God. It either acknowledges God or rejects Him. But math cannot be neutral. If we claim to view math lessons neutrally, we blind our heart to God's presence in math and unconsciously swallow unbiblical, independent thinking.

Chapter 5:

Harm to the Heart

We are so accustomed to viewing math as a neutral subject that I would like to spend a little more time showing the danger of this viewpoint. Let me share with you a few of the ways my belief in math's neutrality harmed me for many years.

Because I believed in math's neutrality, I blindly believed everything my math textbooks told me. I did not look at math with discernment like I did my other subjects. While in other subjects I questioned my textbooks and checked out their teaching against Scripture, I simply accepted everything my math book said as infallible.

It was very clear to me that math was reliable and always worked. While math's consistency should have reminded me to trust God, it instead reminded me to trust math and my math book.

When I entered my teenage years, I began to struggle to make my faith my own. My view of math hampered me in this process. I knew that evolutionists claimed they had mathematically proven the earth was millions of years old. If math always worked, should I not believe what these highly learned mathematicians told me?

If I had known the biblical truth about math, I would have seen math as a tool to record God's orderly world, not as an infallible means to determine the earth's origins. I would have known that math's very existence, in fact, testifies to the fact that an awesome Creator, not the random process of evolution, holds this earth together.[1] But because of my belief in math's neutrality, my math studies instead subtly encouraged me to trust math, mathematicians, and human intellect.

Math is like a living monument, urging us to stand in awe of whatever keeps it working so marvelously. Math is a

---

[1] For more information on the creation/evolution issue, visit Answers in Genesis' website, www.answersingenesis.org.

truly amazing and complex subject. Numbers all interrelate consistently and predictably. We can use math, year after year, time after time, and it always works. Whether consciously or unconsciously, we give the credit for math somewhere. Math will always encourage us to trust either God or something else. As we saw in the first two chapters, math should remind us to trust God. We should see math as a record of the incredible consistency with which God sustains all things. We should see God in math. Math should be a monument to His faithfulness and wisdom.

When we view math as something neutral or independent from God, math encourages us to trust math itself and our own human wisdom and reasoning. We no longer see math as a monument to God. Instead, we see math as a monument to man and numbers.

There are also other ways a belief in math's neutrality harms our hearts. I remember the confusion I felt as I entered high school. Math seemed like a big mystery to me. Why did math work? Where did math come from? My textbooks never really told me. I found myself memorizing rule after rule without really understanding how that rule came about. I began losing sight of the purpose behind learning math. How would I ever use exponents and algebraic division/graphing in my own life?

Had I understood that math was not neutral—that math merely records the order God created—math would not have been a mystery to me. I would have realized that the rules in math are merely ways of writing the complex principles by which God holds everything in the physical universe in place. Had my textbooks presented math biblically, they would have taught me how exponents and algebraic division/graphing are useful, God-given tools. They would have shown me that math has a meaning and a purpose.

Sadly, many students today fail to see math's purpose. They do not realize that math is only a way of recording real-life principles. Instead, they wonder, "Why do I have to study math?"

In short, viewing math as a neutral subject causes us to exchange the truth of God for a lie. It causes us to unconsciously adopt independent thinking and rely on man's capabilities and intellect.

Chapter 1 of Romans reminds us that God's creation testifies to Him. We have seen that math is no exception to this. Each and every part of math testifies to God's faithfulness, greatness, and care.

> For since the creation of the world God's invisible qualities—his eternal power and divine nature—have been clearly seen, being understood from what has been made, so that men are without excuse.
> (Romans 1:20)

However, a few verses later God warns us that we would ignore Him and exchange His truth for a lie.

> They exchanged the truth of God for a lie, and worshiped and served created things rather than the Creator—who is forever praised. Amen.
> (Romans 1:25)

✦✦✦✦✦✦✦

Math is not a neutral subject. God's invisible qualities are clearly seen in math. If we are not worshiping God in math, we are unconsciously worshipping the creation, in this case, math itself. We are exchanging God's truth for a lie. Our attempts at neutrality only harm us and open up our hearts to worldly, independent thinking.

Chapter 6:

Adopting a New Heart toward Math

If the typical way of teaching math harms the heart, how, then, *should* we teach math? How can we teach math biblically?

I used to think teaching math biblically meant adding a Bible verse or two to a math lesson. If we start our math lessons with a Bible story or Scriptural truth, will we have taught math biblically?

Let us try it. Below is one of the lessons we looked at previously.

> Searching for a missing factor is called **division**. A division problem is like a miniature multiplication table. The product is inside the box. The two factors are outside the box. One factor is in front and the other is on top.[1]

Now we will add the following Scripture verse and thought to the lesson.

> For the word of God is living and active. Sharper than any double-edged sword, it penetrates even to **dividing** soul and spirit.
> (Hebrews 4:12)

God's Word is powerful and can **divide** our very soul.

Did we make the lesson biblical? We brought in a truth from God's Word, but we did not change the way division itself is presented. The student is not told that God created and

---
[1] *Hake and Saxon, Math 65, 74.*

sustains division. Rather, he is left viewing division as something independent from God—as a self-existent or man-made fact. This lesson still presents math as something independent from God. Remember, math will always be taught in one of two ways:

- Dependent on God, i.e., something He created and sustains; or
- Independent from God.

Although including Bible verses and thoughts might seem like a good idea, these inclusions alone do not teach math biblically. They may encourage the student to depend on God and His Word for his spiritual needs, but they do not teach him to depend on God and His Word in *math*. They do not portray math itself as something dependent on God. Adding a Bible verse or thought to a typical math lesson still teaches math the world's way. It still depicts math as an independent fact. The student is still taught to think about math as something independent from God. Dangerous, independent seeds are still being planted.

Below is an example from a Christian textbook of trying to add the Bible to what is still ultimately an independent presentation of math.

Thought: Just as God is our Foundation, so real numbers are the foundation of algebra.[2]

Can you see the harm in this thought? Without intending to, this thought actually reinforces the wrong view of math. It implies that God is our foundation, but real numbers, not God, are the foundation of algebra. The student is taught to look at algebra as something independent from God when he should be taught that algebra is dependent on God. Far from revealing God in math, this sort of teaching actually encourages the student to trust God in every area *but* math.

Simply adding Bible verses or thoughts to our math lessons does not make math biblical. Math still comes across as independent from God.

Here is an illustration. If you saw a bomb that was ready to explode, what would you do? Would you tape Bible verses over

---

[2] *Accelerated Christian Education, Algebra II, 1995 rev.* (USA: Accelerated Christian Education, 1993), pace 1121, 10.

the bomb until you could no longer see the danger? Of course not! That would be foolish. You would seek to diffuse the bomb.

In math, independent thinking is like a bomb ready to explode. Adding Bible verses to our math lessons will not change the dangerous thinking. The danger can only be "diffused" by rejecting the independent thinking and embracing a biblical, dependent view of math.

While merely adding Scripture verses to math encourages students to build their life biblically, it still teaches them to build their thinking about math on a worldly, independent philosophy. To visualize the foolishness of this, picture a house being built. Can you imagine a builder laying a firm foundation on which to build most of the house, but deciding to build a window on the nearby sand? How foolish! Is it not equally foolish to encourage students to build their life on Christ, the firm foundation, while at the same time teaching them to build their view of math on the sands of worldly thinking? How much better to encourage them to build *every* part of their life, including their view of math, on Christ!

Return to our sample lesson on division.

> Searching for a missing factor is called **division**. A division problem is like a miniature multiplication table. The product is inside the box. The two factors are outside the box. One factor is in front and the other is on top.[3]

---

[3] Hake and Saxon, Math 65, 74.

This time, instead of adding Bible verses to the lesson, we will try changing the heart in the presentation. Rather than introducing division as an independent fact, we will present division as a practical way of recording a principle God created and sustains. Assuming we are just teaching division and not the topic of missing factors, our division lesson might resemble the following:

> Today we are going to learn about division. Suppose you have twelve cookies that you want to share evenly amongst three friends. How many cookies should you give each friend? Try it! Take a piece of paper and cut it into twelve circles to represent the cookies. Now split the cookies (circles) into three equal piles. How many cookies are in each pile? Four. You just did division! You split, or divided, a big number (twelve cookies) into a certain number of piles (three) and found out how many went into each pile (four). Guess what? Because God holds all of creation together so consistently, if you take twelve objects and put them into three piles, you will always end up with four objects in each pile! God's consistency makes it possible for us to do division without actually touching the objects we are dividing. Just like we did with multiplication, we can memorize our division facts and be confident that they will always work because of God's faithfulness.
>
> Now we will learn one way mathematicians have agreed to write division problems. They have agreed to put . . .

The above lesson presents division as a useful, real-life tool that is dependent on God's faithfulness. It introduces this method of writing division as just *one* way mathematicians have agreed to record division. It teaches division as a principle dependent on God.

The above example is just one example of how division can be presented biblically. As we will explore in depth later, there are many different ways to teach each math concept biblically. But a biblical math presentation will *always* introduce the concept as a practical tool dependent on God's continual faithfulness. When we approach and teach each concept in this biblical light, math will stop harming our hearts and start reminding us that we can trust God.

Isaiah 55:8 reads: "'For my thoughts are not your thoughts, neither are your ways my ways,' declares the LORD." In math, as in every other area of life, we have the privilege of laying aside our old thoughts and receiving God's thoughts. How foolish it would be for us to add the Bible to our old thoughts about math! Rather, may we adopt God's thoughts in our math outlook and presentations.

Math Methods from the Past

As you may have already noted, teaching math biblically involves a practical element. We want students to see math as a method of recording the way objects interact in the physical world so that they will realize that math is dependent on God's sustaining hand. Since most of us are so accustomed to viewing math as a paper exercise, the idea of teaching it practically seems foreign to us. I know I used to wonder how math could possibly be taught without endless drills and memorization.

You may be as surprised as I was to find out that math has not always been taught in the current fashion. Two hundred years ago, most textbooks taught math in an incredibly practical way. Look briefly with me at how math used to be presented. Seeing these historic math methods will show you that math does *not* have to be taught through rote drills. Math can be taught as a meaningful tool.

Arithmetic Then…

The *Scholars Arithmetic* was a widely used textbook in the early 1800s.[4] The approach implemented in this book can best be termed practical—practical to a degree I had never imagined possible. Anyone reading this book will feel like they are

---

[4] Educational Research Library (National Institute of Education), *Fifteenth to eighteenth century rare books on education: a catalog of the titles held by the Educational Research Library* (Washington: National Institute of Education: For sale by the Supt. Of Docs., U.S. Govt. Print. Off., 1976), 201.

reading a here-are-the-tools-you-need and here-is-how-to-use-them handbook instead of a math textbook. As the author says:

> The *thing needful*, and that which distinguishes the arithmetician, is to know how to proceed by application of these four rules to the solution of any arithmetical question. To afford the scholar this knowledge is the object of all succeeding rules.[5]

"Any arithmetical question" meant not just problems in a math book, but any *real-life scenario* the scholar might encounter. This book taught math as a tool for real life. Using math practically was presented as the main purpose of learning math, not just as an afterthought.

Imagine learning math, not because you had to, but because each and every concept you learned would help you in your daily life. Imagine learning along with subtraction all of the possible uses of subtraction including: currency conversion, bookkeeping, and even basic mechanics. Basic arithmetic two hundred years ago was exciting, practical, and purposeful. History demonstrates that arithmetic can be taught as an exciting, practical tool.[6]

Geometry Then...

Two hundred years ago, geometry was not the pile of proofs it has become today. Geometry in the 1700s remained true to its technical Latin meaning, "earth measure." Geometry served as a tool to help men measure and record the earth.

Goedaesia, the geometry book of the time, was actually a surveying manual. Geometric rules did not just help the student prove an irrelevant fact. They helped him find the height of a tree, the area of a field, or the circumference of a lake. Shapes represented real objects and served a real purpose.

---

[5] Daniel Adams, *The Scholars Arithmetic; or, Federal Accountant. Containing . . . The Whole in A Form and Method Altogether New, for the Ease of the Master and the Greater Progress of the Scholar*, 5th ed. (Leominster, MA: Adams & Wilder, 1804), 57.

[6] I do not mean to imply that textbooks two hundred years ago were perfect. They were not. They had weaknesses. However, they presented each math rule as an incredibly practical tool the student could apply to his own life. Their practical approach serves as a good example.

Men in the 1700s used geometry on a regular basis. At the time, few skills were more valuable than surveying. Young men, many much younger than today's typical geometry student, taught themselves geometry so that they could survey property. Knowing the shape and features of the land was vital to protecting, expanding, and using property wisely. Since most men designed their own homes and gardens, knowing how to survey was an essential skill. Students learned geometry so they could use it in their own lives.

♦♦♦♦♦♦♦

Math textbooks from the 19th century illustrate that we do not have to teach math in a meaningless fashion. Math has been taught practically in the past and can be taught practically today. It *is* possible to entirely change the way we teach math.

NOTE

Just presenting math practically does not teach math biblically. What a tragedy it would be to know how to use math practically while remaining blind to the One who makes practical math possible! Yet practical math is the natural outcome of a biblical view of math.

The Joy of Biblical Math

Not only is it possible to change the way we view and teach math, but a biblical approach to math is also very rewarding. Briefly journey further back in history with me and take a look at two very different groups of men: the Classical Greek mathematicians (circa 600-300 BC) and the men of the Scientific Revolution (circa 1500-1600 AD). These two groups approached math in radically different ways. Their comparison demonstrates that our thinking about math directs our actions, and that biblical thinking in math brings rewards and immense joy.

We will start with the Classical Greek mathematicians—men like Pythagoras, Plato, Aristotle, and Euclid. These mathematicians believed that logical, mathematical reasoning could determine anything. They placed their entire faith in human intellect and math. They gave the credit for math's incredible consistency to man and math itself. They viewed math as something totally independent from God.

What were the results of this independent perspective of math? The Greek mathematicians, while they made many contributions to math, were never able to fully understand math and why it worked. Instead of using math to develop practical inventions, they spent most of their time trying to explain how the universe originated and operates using math and human reason alone. Math to them was an infallible reference point, an intellectual exercise, and a testimony to man's genius.

To see the full impact of this Greek philosophy, we need to fast-forward in time to the early Christian Era (300-600 AD). During this time, the church began compromising with Greek teaching. The Greek philosophy and approach to math began to infiltrate European culture, leading to many inaccurate ideas and hampering scientific progress.

For example, Greek teaching led to the misguided belief that the sun circled the earth. The Greeks had used math to "prove" that the sun circled the earth. Due to their faith in math and human intellect, the Greeks did not verify this proof with science. Instead of questioning the Greek model of the universe, Europeans twisted Scripture to support this false view.[7] Why? Because Europeans embraced the Greek faith in math's infallibility and man's reasoning. The Greek approach to math and science held Europe in an intellectual and spiritual vise that restricted academic progress and bound society to false ideas.

---

[7] In an effort to make a geocentric philosophy fit into the Bible, the Medieval church claimed that the sun standing still in Joshua 10:13 meant that the sun circled the earth. However, at the time Scripture was written it was commonly known that the earth circled the sun. Scripture merely uses the earth as a point of reference, which is actually the most scientifically correct method of referring to heavenly movements. See Henry Morris with Henry Morris III, Many Infallible Proofs: Practical and Useful Evidences for the Christian Faith (AZ: Master Books, 1996), 253. Quoted in Russell Grigg, "Joshua's Long Day," Creation, June 1997, http://www.answersingenesis.org/creation/v19/i3/longday.asp.

Moving forward in history, we encounter a radically different group of men: the men of the Scientific Reformation. These men emerged on the scene toward the end of the Middle Ages, a period of time when, despite the corruption of the established church, God's principles began correcting and replacing independent Greek thinking. Most of these men were heavily influenced by the Protestant Reformation, a reformation that called men back to the Bible as the basis for every area of life.

Instead of seeing math as an intellectual pursuit of truth, the men of the Scientific Reformation viewed math as a way of recording the truth God had already created and sustains. They embraced math as a vital part of science. They acknowledged their utter dependence on God and recognized Him, not math, as their ultimate authority.

No longer blindly trusting math and man, men like Galileo Galilei and Johannes Kepler did not just accept the "proven" geocentric view of the universe. Rather, they questioned the Greek proofs, knowing that even the best-laid mathematical proofs could have flaws. Instead of placing their faith in mathematical reasoning, they placed their faith in the Bible. Learning from the sacred pages where the universe originated, these men wasted no

time in idle speculation. Instead, they looked in awe at the heavens God stretched out above them and used math to record what they saw. By so doing, they discovered numerous laws (such as the laws of planetary motion) and realized that, contrary to the Greek proofs, the earth circles the sun.

NOTE

Seeing how a biblical view of math helped bring about the incredible scientific discoveries of the Scientific Reformation should encourage us that the Bible's principles work. We can safely build our understanding of math on God's Word.

Besides just being able to discover previously unknown facts about the universe, these mathematicians/scientists experienced the joy of seeing God's hand as they used math. They knew beyond a shadow of a doubt that math and creation were the Creator's masterpieces and mere reflections of His power. They knew math worked because God faithfully sustains the universe.

For example, Johannes Kepler, the discoverer of the famous laws of planetary motion, clearly acknowledged that the intricate workings of the heavens testified to the Creator's hand. In fact, he realized that God had actually hung the planets in such a way that they produced a sound. Kepler recognized that math merely shows us the design and sounds that would otherwise be hidden from our sight. His biographer states that to Kepler, "God is the beginning and end of scientific research and striving."[8] Regarding the laws of the planets, the same biographer comments that Kepler believed that, "The eccentricities of these ellipses are no more arbitrary and without rule than any other measures. No, in this fine construction the highly artistic formative hand of the Creator is shown in a very special way."[9]

Kepler knew why the heavens were orderly. He understood why math worked. The Bible gave him the answers. He

---

[8] *Max Caspar, Kepler, trans. C. Doris Hellman (New York: Dover Publications, [1959] 1993), 374, quoted in Nickel, 117.*
[9] *Ibid., 386.*

Adopting a New Heart toward Math

viewed each scientific and mathematical endeavor as a chance to see more of God's hidden design. He approached math with a heart to see God, and as a result, he was able to make startling discoveries about the universe.

Like Kepler, other men of the Scientific Revolution saw God in math because they thought about math biblically. A total heart change enabled them to view and use math appropriately and effectively.[10]

♦♦♦♦♦♦♦

If you want your children to see and use math biblically, I would encourage you to do more than just add Bible verses to your curriculum. Let God change your heart toward math. Like Kepler and the other 16th and 17th century mathematicians who rejected the independent Greek thinking, reject the current independent thinking in today's "neutral" math books. Adopt biblical thinking.

Embracing a biblical philosophy toward math will enable you to use math practically in such a way that math will be a monument in your hearts to God's greatness. As you begin to see and use math biblically yourself, you will be able to teach math biblically to your children so that they, too, can behold God in math.

---

[10] *For more information about mathematicians throughout history and their belief systems, see Nickel, Mathematics: Is God Silent?*

Chapter 7:
Preparing to Teach Math Biblically

Like anything else in our Christian walk, teaching math biblically will take strength and endurance. It is unnatural for us. We are used to viewing math independently. For that matter, we are used to living independently from God. Drawing our strength from Him and looking at things His way runs contrary to our sinful nature. The Bible says we have to *train* ourselves for godliness (1 Timothy 4:7). Training implies effort. The rest of this chapter will provide a few training tips.

Remind Yourself of God's Truths

Continually reminding yourself of God's truths will help you keep on course. "Stand firm then, with the belt of truth buckled around your waist" (Ephesians 6:14a).

Below are a few key truths to keep in mind as you teach.

- God created math.
- Math works because of God's faithfulness in sustaining all things.
- Math helps us explore and utilize creation. As we use math, it should also make us praise the great Creator.
- Math is not a neutral subject.
- Math was designed to bring glory to God.

If you want to be prepared to teach math biblically, continually remind yourself of God's truths. Surround yourself with as many biblical resources as possible. While there are not many biblical math resources on the market, biblical resources on other subjects and input from other Christians can help you keep focused. By whatever means you can, keep God's truths continually before you.

Examine Your Goals

The Bible tells us David's heart was, "fully devoted to the LORD his God" (1 Kings 11:4). We constantly need to ask ourselves, is our heart fully devoted to God in math?

Many times we devote ourselves to our own goals and desires when we teach. We want to be good teachers. We want our students to succeed academically. We strive for these things and devote our time and energy in these directions. We worry unduly over our students' academic progress.

But stop and think for a moment. What is God's goal for your child?

From the Bible, we learn that God's goal for all His children is for them to walk and live in His Spirit, leaning on Him. The *Shorter Catechism* declares that, "Man's chief end is to glorify God and enjoy him forever."[1]

The world defines success as good grades and intellectual knowledge, but God does not. God's primary goal in math, as in everything else, is for us to worship and learn from Him. Success in His eyes is seeing and trusting Him. You can teach your children to be proficient in math, but if they have not learned to glorify God in the process, they have fallen short of His goal.

---

[1] *The Shorter Catechism, agreed upon by the Reverend Assembly of Divines at Westminster*, quoted in *The New England Primer, Improved.... (Boston: printed by Edward Draper, Newbury Street and sold by John Boyle, Marlborough Street, 1777;* reprint ed., Aledo, TX: WallBuilder Press, 2003), Question 1.

I would like to encourage you to make God's goal your goal in math. Strive to teach your children each math concept biblically. Whether they can manipulate numbers and quote rules should be of secondary importance. More importantly, do they realize how each math concept points to God? Do they know how to use it for His glory? Do they have a firm foundation and perspective in math? Are they applying that perspective to each new concept?

A thousand other goals will continually try to make their way into your heart. When my mom first started homeschooling me, her goal was for me to love God and read His Word. Over the years, however, God began to show her that she had many other hidden goals for me. She wanted me to be everything she had longed to be herself. She wanted me to succeed and prosper. These other goals and desires often hampered her teaching and kept her from focusing on God's goal. She found she needed to repeatedly surrender and ask God to truly make Him her only goal.

As the Lord shows you other goals you have for your children in math, surrender those goals to Him. Continually ask God to make Him your only goal.

♦♦♦♦♦♦♦

The first step to teaching math biblically, and one to which we need to constantly return, is to make sure our own hearts are truly prepared to worship God in math. We need to constantly keep God's truths and goals in the forefront of our hearts and minds.

Chapter 8:
Teaching Math Biblically

Every person is so unique and different that no two parents could possibly teach math the exact same way. The way you teach math will be uniquely yours. You may or may not be naturally gifted in math. You may or may not have the same ideas and teaching style I do. You may or may not have time to prepare. Your time with your child in math may be short or long. You may or may not use a textbook. And that is okay! God knows your makeup and limitations. He custom-designed you for your child. So please do not take the ideas I share here as rigid rules to follow. View them instead as guidelines and principles. Let them be a starting point for you, not a formula to follow precisely. Allow God to teach you and give you other ideas.

General Guidelines

While there is no one formula for teaching math biblically, some universal principles always apply. Biblical teaching will always teach the child to view math biblically and use it practically.

To accomplish this, you will need to do more than simply say to your child "God created math" or "Math is practical." You will need to show your child how each concept testifies to God and teach him how to use it practically. Suppose you did not know how to drive a car. You might need someone to verbally explain how to drive, but eventually you will need someone to put you behind the wheel and actually teach you how to drive! Likewise, while you should explain to your child that God is in math and that math is useable, do not stop there. Make sure that you are actually teaching your child to view each concept biblically and use it practically.

Teach your child when and how to use each concept.

Mathematics is like a chest of tools: before studying the tools in detail, a good workman should know the object of each, when it is used, how it is used, what it is used for.[1]

(Walter W. Sawyer)

Students should know and understand math like a workman understands his tools. They should understand where each tool (math concept) came from and why it works. Then they should learn why and how to use that tool (math concept).

In short, whenever you teach math, strive to:

- Show the student how the concept reveals God's character/design; and
- Teach them to use the concept practically.

---

[1] Walter W. Sawyer, *Mathematician's Delight* (Harmondsworth Middlesex: Penguin, 1943), 10, quoted in Nickel, *Mathematics: Is God Silent?* 290.

To help convey these ideas, you will frequently want to incorporate:

- The history of math; and
- The practicality of math.

You may have observed in previous chapters that math's history and practicality help reinforce a biblical view of math. Seeing the way men throughout history have used math reinforces math's usefulness and connection with reality. Looking at the different approaches men have taken to math emphasizes the importance of our beliefs about math. In these and numerous other ways, understanding math's history and practicality helps us view math correctly as a useful, God-given tool. Therefore, you will want to consistently include these elements in your teaching.

With this general overview in the back of your mind, let us look in more depth at specific ways you can teach your child to view math biblically and use math practically.

NOTE

While some of the examples below might not be age-appropriate for your child, you can modify them to fit your child's math level. It is never too early or too late to begin teaching math biblically!

View Math Biblically

How can you teach your child to view math biblically? As mentioned in previous chapters, start by approaching math with a heart dependent on the Lord and His Word. Then, look at every concept you teach and ask God to show you how He views it. Much like we did with the equation "1 + 1 = 2" in the first chapter, strive to present the concept in a biblical way. To illustrate, I have listed a few general math concepts below and summarized how each concept can be presented biblically.

Counting—Counting should be introduced as an extremely useful tool that works because of God's faithfulness in holding all things together. In a very basic way, counting helps us quantify the order God has placed around us. For example, count your fingers. Now count someone else's fingers. Of course, you both have five fingers on each hand. Although we do not often think about it in this light, everyone has five fingers because the same God created us all, and He decided that hands would have five fingers.[2] Counting should make us praise our Creator.

As you teach counting, explain to your child that the Bible also tells us why we can count, or for that matter, use math at all. We, unlike the animals, can count because God created us in His image, capable of seeing and recording the order around us.[3]

Division—When you teach division (or other basic concepts), remember why division works (because God holds all things together). Present the many different ways of writing division problems ($y\overline{)x}$, $\div$, $\frac{x}{y}$) as different ways men have agreed to record the fixed law of division God has established and sustains.

Algebra—Since algebra integrates numbers and letters, stress that every letter represents a real-life object. The letter *a* could represent age, apples, air, atmospheres, or any other unknown value. The letter *c* might represent a can of beans, carrots, the current year, currency, or any other unknown value. By using letters instead of numbers, we can write general relationships (such as $a^2 + b^2 = c^2$) that hold true no matter what numbers are inserted into the equation.

Obviously, algebra is a helpful tool because it enables us to describe, not just one particular situation, but the way things will interact in *any* situation. Why do things interact so consistently?

---

[2] *Exceptions to this exist because of the fall of man. The world is no longer perfect like God created it, although we can still see remnants of its original order.*

[3] *Animals were not created in God's image. Thus it is impossible for them to reason like man can. For more information on this or other issues dealing with creation, visit www.answersingenesis.org.*

Because God faithfully holds creation together in a consistent manner. Algebra is a beautiful testimony to God's faithfulness and power!

Geometry—Since the Classical Greek mathematicians contributed so much to geometry, you will have to be especially leery of independent thinking in geometry presentations. Guard against teaching geometry and postulates/theorems as independent truths. Postulates/theorems only help us explain and draw creation, enabling us to find the height of mountains or trees from the ground with very little effort.[4] Geometry helps us lay out gardens, create maps, build houses and machines, and graph complex statistical data. Make sure you present geometry as a useful tool.

Use Math Practically

Telling a child the biblical view of a concept is one thing; showing him how to use the concept practically is quite another. As you teach, you not only want to bring a biblical view of math, but you also want to make sure your child actually learns to use the concept practically.

Lay aside the textbook as often as possible and have your child apply math to real-life situations. Beyond just helping him learn math practically, using math in his own life will reinforce for your child that math, far from being a man-made system, is a useful way of recording and utilizing principles God created and sustains. Real-life situations help emphasize the fact that math is more than a mere intellectual exercise.

There are many ways that you can incorporate real-life situations into your math teaching. Below are a few examples. While the examples are geared toward the younger grade levels, the same principles apply in the older grades.

Sorting wash or setting the table can demonstrate the function of sets. You can have your child learn and apply basic addition/subtraction or multiplication/division with real objects, such as pencils, toy cars, dry elbow noodles, beans,

---

[4] Nickel, *Mathematics: Is God Silent?* 47-48.

candy, or anything else you have on hand. While teaching graphs or measurements, you can have your child measure his different body parts (such as his arms, wrists, and legs) and cut a piece of ribbon the size of each part. Then have him use the ribbon to graph and compare the sizes of his different body parts. When presenting decimals, you can open a play store and let your child operate the checkouts (he will not even guess that he is learning math). You can incorporate real life into fractions by having your child double or half a recipe with you, or by going on a made-up treasure hunt around the house and dividing the spoils equally.

These examples are just a few of the thousands of simple and fun ways you can teach math without a textbook. Remember, bringing math into real life can help you show your child how math, far from being an isolated, intellectual subject, is a useful, God-given tool.

Sometimes reinforcing a concept with a real-life example is not enough. It is often better to demonstrate the concept in the physical world *before* showing the child how to write and work with it in the textbook. This teaches him to think mathematically and gives him a proper perspective on the concept when he later learns it in his textbook. Above all, it reinforces that math is a way of describing the physical world.

Instead of always presenting concepts to your child, try to occasionally let him discover a concept on his own (with your guidance, of course). Let your child investigate real-life problems and learn to use math to solve those problems first; then teach him the textbook mechanics of the concept.

For example, instead of reading to your child about sequences, look at several flowers, fruits, or seed heads that contain sequences (such as sunflowers—you can consult a reference book to find out what flowers or materials to have on hand). Then guide your child into seeing for himself the unique sequences God used in creation. You have now taught him how to explore God's world with math.

Suppose you are teaching rounding. Instead of reading a textbook, take your child to the grocery store and tell him to keep track of all the items you put in the cart. Your child will quickly realize that there is no way he can ever keep track of

Teaching Math Biblically 61

$1.23 + $2.99 + $2.50. Gently guide him into discovering rounding for himself as you shop. Not only have you taught him rounding, but you have reinforced in his mind math's usefulness. He has learned how to round numbers in a real situation *before* he ever learned rounding in a textbook.

One mom shared with me how she adapted this idea to fit her family. She told her oldest daughter she would buy her and all her younger siblings a piece of candy if she could estimate their groceries for the week within two dollars. With the prospect of candy as an incentive, the daughter paid careful attention the entire time they shopped and actually estimated the total within a few cents. Amazed, the daughter later remarked, "Wow, Mom, math really *does* work!" The younger children, who had watched the whole proceeding with a vested interest, also learned that math was fun and useful.

Teaching math from real-life situations works. It helps the child view math correctly as a God-given, practical tool. While I do not mean to imply that you will *always* teach without a textbook, I would like to encourage you to break from the idea that you must *always* teach from a textbook. Since we want to see math as a real-life tool, we need to remember to use and teach math from real-life situations.

Putting It All Together

To give you a better idea how the above guidelines impact the way you teach, I would like to walk through teaching two concepts—lines and fractions. First assume that your textbook is presenting the definition of a line. It will probably say something like, "a line is a never-ending series of dots," and then move on to defining line segments or number lines. How can you apply the principles mentioned above to this lesson so that your child will view lines biblically and use them practically?

You could rework a lesson on lines in many different ways. How you choose to present lines will depend upon the ideas the Lord has given you at the time. Some ways might require research on your part, and the next chapter will explain how to easily find the information and ideas you need.

One way to teach lines would be to revise the textbook's independent presentation so that it presents lines biblically. For example, instead of saying, "a line is a never-ending series of dots," you could explain that a man named Euclid decided to name a never-ending series of dots a line. This little change helps present a line as a name chosen by a mathematician to express a real-life concept, whereas the original wording gives the impression that a line is some sort of self-existent fact.

Or you could use lines to discuss the vastness of our universe with your child. To us, lines are a never-ending series of dots. We could not even begin to draw a line that would reach the end of our universe. Yet God holds the whole universe in His hands. He knows its end and beginning. Lines are a beautiful testimony to God's mightiness! Since a ruler is really a portion of a number line (or a line segment), you could incorporate measuring into the lesson. Select a project that involves measuring so that your child can use number lines himself.

Lines also provide a great opportunity for you to discuss the Greek mathematicians. You could talk about how the Greeks failed to see math biblically and the harm this caused.

If you have time to do a little more research, you could look up lines in a reference book. There you will discover that many other definitions for a line exist. Most textbooks define a line as a never-ending series of dots. This is the definition of a line in Euclidian geometry. Riemannian geometry, on the other hand, gives a line an entirely different and contradictory definition.[5] Both definitions are useful in different situations.

Introducing these contradictory definitions to your child is a wonderful way to teach lines. You could explain that, just like forks and spoons help us eat different foods, different geometries help us in different situations. A fork has its limits on what it can help us eat (it does not do well with soup), and geometry definitions have limits on what they can help us record. God has created a complex universe, and each of the geometries helps us explore and record a portion of it. None of

---

[5] *Riemannian geometry, in simple terms, defines a line as a "great circle." This would mean a line has a beginning and end and that the equator is really a line, since it has a beginning and an end. See Nickel, Mathematics: Is God Silent? 178-179.*

the geometries contain perfect definitions—they are merely tools that must be used appropriately to be accurate.

Another way you could teach lines, especially to older children, involves setting aside the textbook entirely. Instead of reading the textbook lesson, you could have your child research lines in other math books and report to you on their meaning, history, and uses. This way, your child will remember the concept *and* learn to view and use it appropriately.

As you can see, lines can be taught in many different ways. Notice that each way reinforces a biblical understanding of math. Much like there are many different ways to explain or illustrate the gospel message without changing the message itself, there are many different ways to teach math concepts. Yet these ways should *always* reinforce a biblical understanding of math.

Now suppose your textbook is teaching fractions. It will probably go into depth about numerators and denominators, what they mean, and the proper way to write them. After briefly introducing the mechanics of fractions, most curriculums make students solve numerous problems. Instead of merely making your child solve a host of rote problems, how can you show your child a biblical view of fractions and teach them to use fractions practically?

To help your child develop the correct view of fractions and the way to write them, you could explore with him how different cultures have written fractions (for instance, the Egyptians wrote $1/2$ as $\frac{\frown}{11}$).[6] Seeing other ways of writing fractions will help your child understand that our current way of writing fractions, much like our current language system, is just one way of expressing a real-life concept.

You could also teach fractions by incorporating some science into your math lesson. Lots of simple science experiments involve measuring or weighing fractional amounts. Performing a science experiment can relieve the boredom of studying fractions and demonstrate to your child many practical uses.

---

[6] Adler, *Giant Golden Book*, 43.

On other days, you might use the textbook a little more. Suppose you teach fractions using Egyptian history one day. The next day, assuming your child understands fractions biblically, you might decide to let him read and work problems in his textbook for math class. Then, to reinforce that numerators and denominators represent real things, you might have him cut the pie for dinner that night into eighths, or divide objects around the house into fractions with you.

♦♦♦♦♦♦♦

Do you see how the above ways of teaching both lines and fractions followed the guidelines we discussed earlier? They presented lines and fractions from a biblical view. Many of them provided opportunities for the child to use the concept away from the textbook. Others brought history and science into the lesson.

As you teach math, strive to use whatever concept you are teaching as an opportunity to show your child the truth about math and to teach him to use math appropriately. This, in summary, is what teaching math biblically is all about.

Chapter 9:
Ready, Set, Now What Do I Do?

You sit down to plan your child's math lesson. You open the book and stare blankly at the page. You have no idea how this concept should be taught biblically. What do you do? How do you find ideas?

Below are a few suggestions to help you when you feel at a loss. First, write down what concept/concepts you will be teaching to your child (sometimes this simply involves copying the Table of Contents). Then ask yourself the following questions:

- How does this concept testify to God's faithfulness or power?

- Why is it important to learn this concept and/or what can be done with the information learned?

- How has this concept been used in the past? How is this concept used today? Did the men who discovered this concept use math correctly? Why or why not?

These questions will help you see how the concept points to God and serves as a practical tool.[1] If you do not know the answers to the questions you ask, do not worry. Sometimes God's hand in a concept is not obvious. Oftentimes you may not know why the concept is important to learn or how it should be used. You frequently may not know a thing about the concept's history. But that does not mean you have to panic. It simply means you have to dig a little deeper.

---

[1] Notice how these questions directly relate to the general guidelines for teaching math on page 55-57. These questions help us think of how the concept reveals God's character/design, applies practically, and relates to science/history.

View math as a treasure hunt. Sometimes a treasure is easy to find; other times, it takes a little more effort. Likewise, ways to teach a concept biblically will sometimes be obvious to you, while at other times you will have to hunt for the answers you need. Much like a miner who spends his whole life digging for gold, embark on a life-long discovery trip. If you are constantly on the lookout for ways that math points to God and is useful, you will find lots of ideas for teaching.

You may find it helpful to create an idea notebook to aid you while you teach. Use this notebook to jot down all the practical uses for math you discover. If you find yourself using math to measure fabric for a dress, write that down in your notebook. If you read about a new scientific discovery in the newspaper that used math, make a note of it. Then on those days when you need ideas, you can use your notebook for inspiration.

In Appendix A, you will find a sample notebook. The notebook lists a few common ways almost all of us use math, as well as some basic ways you will likely encounter math in science. The examples listed there are just to get you started. The rest of this chapter will show you how to begin to find your own ideas for math in everything you do.

In Your Curriculum

Start hunting for ideas in the math textbook you are using. Check to see if it gives a clue on how the concept was discovered or used. If it does, take the examples it offers and explore them in more depth with your child.

In Resources

It is often helpful to consult other resources besides your math textbook. Several good resources are listed on pages 76-77. You may want to have a few of these resources on hand.

Also, take advantage of the resources at your local library. For instance, if you need graphing ideas you should easily be able to locate a helpful book on graphing in your library's junior section. Look for books that sound practical or even have key words such as "practical" or "real life" in their title or description. I have found several useful books by just scanning my local library's bookshelves.

The Internet is also a great information source. Frequently, you can find out about a math concept simply by typing its name into a search engine. For example, I found details about the incredible way seeds are arranged in sunflowers by searching for "Fibonacci sequence" online (I found out about the existence of this sequence initially by looking up "sequences" in the index of *Mathematics: Is God Silent?*, one of the reference books listed in the next chapter).

As people share with me ways they have applied the principles in this guide, I will post their thoughts along with any further ideas or suggestions on my website, www.christianperspective.net, so you may want to check the site periodically. You will also be able to email your questions and share ideas there.

In Your Own Life

For other ideas, think about how you have used the concept in the past. We all use math a lot more than we realize. Many times, the ways you have used a concept will give you ideas on how to teach it as a practical tool.

Keep your eyes open for math in other subjects. Science and math are closely connected. See if math was used to help discover what you are learning in science. For example, while studying earthquakes, research how math is used to record seismic activity. You will generally discover practical uses for several math concepts in each topic you research.

History provides a wonderful opportunity for you to explore math's past uses. For example, math enabled the Egyptians to build gigantic pyramids. So when you study Egypt in history, why not have your child build a sugar cube pyramid?[2]

---

[2] *Instructions on building sugarcube pyramids can be found in Cindy A. Littlefield, Real-World Math for Hands-On Fun! (Charlotte, VT: Williamson Publishing, 2001).*

Use the opportunity to show him the math behind pyramids while reinforcing in his mind math's usefulness and purpose.

When studying Greece, research how the Greek philosophers viewed math inappropriately. As you teach the Protestant Reformation period, explore the lives and math of men such as Johannes Kepler or Sir Isaac Newton. Do not just read about history-making inventions—study the inventions and see how the inventor put math to a practical use. Each and every time period in history can help you teach math.

Music class also provides wonderful lessons on math. Good music follows an orderly and mathematical structure. Whenever you listen to music, pay attention to the song's order or disorder. In the eighteenth and nineteenth centuries, composers built their songs around tones that were mathematically similar. This logical selection of notes created pieces that sounded beautiful and soothing to the ear. However, many musicians in the early twentieth century stopped using notes that went together mathematically. As a result, their music sounds strangely unconnected.[3]

Even art class can turn into a math lesson. After all, good drawings contain proper proportions and shapes. Paints and other mediums must also be mixed in precise mathematical proportions to produce the desired effect.

Nearly every subject and thing you do can give you lots of ideas to use as you teach math, so keep your eyes open.

◆◆◆◆◆◆◆

Wherever you are and whatever you are doing, have a learning attitude. Be ready for God to show you something. Continually hunt for God's ideas in everything you do. Make math a daily treasure hunt. Then you will be ready with plenty of ideas to incorporate into your math lessons.

Also, remember that you do not necessarily need to have each concept researched and understood before you start teaching. You can let your child help you through the process. Math can be a fun treasure hunt for both you and your child.

---

[3] Any guide to classical music should explain the general regularity, or lack thereof, in a piece. Pay close attention to the word, "dissonance," a fancy word for notes not mathematically related to each other.

Chapter 10:
Curriculums and Resources

Guidelines to Follow

There are many math curriculums available. Talk to a group of educators, and each one will tell you what curriculum they feel is the best and cite how well it has worked for them.

Do not be overwhelmed by the choices you have. It is not nearly as important *what* curriculum you use as *how* you use that curriculum. Your heart and approach is much more important than your choice of curriculum. Nevertheless, picking a curriculum can be a daunting task, so I will try to offer a few suggestions.

When choosing a curriculum, keep your goal in mind. Above all, you want your children to see math as God sees it. You want them to understand that math represents principles God created and sustains. You want to demonstrate this through every-day, historical, and scientific examples and exposure.

Ideally, you want to find a curriculum that presents math biblically, but remember that the word "Christian" on a math book does not mean it presents math biblically. There are many "Christian" textbooks that look identical to a secular textbook except for a Scripture verse or saying in the corner. A truly

Christian textbook is one that recognizes God's hand in math and presents math in a biblical and practical way.

Since most Christians believe math is spiritually neutral, there are few biblical math curriculums available. The good news, however, is that there has been a resurgence of practical math curriculums in recent years. The more practical, scientific, and historical information your curriculum contains the better. That way, you will have less to research and add to the program. Look for a curriculum that:

- Makes the student use math practically (preferably not with meaningless word problems but in conjunction with science, accounting, mechanics, or other disciplines);
- Illustrates how concepts are used in science and other disciplines; and
- Gives historical information on concepts.

Factors such as cost, usability, and learning styles may also impact your curriculum choice. Some curriculums, while they might work great for one student, would frustrate another one. Likewise, some curriculums demand more teacher involvement than others. These higher-involvement curriculums, while excellent for some families, might not be the wisest choice for a parent already pressed for time. The curriculum you select will depend in a large measure on your family's specific needs.

In fact, you may find it easier to not use an official curriculum at all. Instead, you may choose to assemble your own curriculum from various resources. There are numerous books on math that, while not actually curriculums, contain a lot more practical and historical information than most curriculums. Do not be afraid to supplement or even replace official curriculums with these books.

If you do opt to use an official curriculum, do not feel obligated to follow the curriculum exactly. View it as a guide to help you, but do not let it stifle you.

NOTE

Just because a resource is listed in the following sections does not mean that I necessarily recommend or endorse it. I have listed a few of the different curriculums currently available, and a handful of those reference/supplemental books that I personally found helpful or thought would be helpful to you due to the mathematical information they contain. I do not agree with all the spiritual content in the listed sources, so please be careful to hold everything up to Scriptural truths.

Curriculums

The curriculums are listed in alphabetical order. Their order or the length of the description does not in any way reflect a preference of one curriculum over another. It is my intent to share a few thoughts on a sampling of curriculums rather than to recommend any one curriculum. I have divided the curriculums into two general categories—manipulative-based curriculums (marked with a ✋ symbol) and textbook-based curriculums (marked with a 📖 symbol). The curriculums that do not fall entirely in either category are listed in the category that best describes them. Here are some general thoughts on these two categories.

✋ Manipulative-based curriculums:

Manipulative-based curriculums teach math mainly through the use of manipulatives. Each math concept is visually demonstrated with manipulatives. Most manipulative-based curriculums also include some textbook work. Because of the manipulative element, these curriculums tend to involve more interaction between the teacher and the student.

If you use a manipulative-based math curriculum, be sure to bring in real-life situations. It is important for the student to use math beyond manipulatives. Additionally, remember to add math's scientific uses and historical roles into the curriculum.

📖 Textbook-based curriculums:

Textbook-based curriculums, while they may occasionally offer hands-on activities, typically teach math mainly through a textbook. These curriculums tend to stress drills and memorization and to give the student a large number of problems to solve every day.

If you choose a textbook curriculum, try to bring nontextbook activities into your math lessons. Be sure to continually bring your child back to the correct heart in math. Use the practical presentations and ideas the curriculum offers as launching pads for further discussion and/or research. It is important for the student to understand that math is not just a textbook exercise. Teaching math from real-life scenarios in addition to the textbook will help your student understand that math is a practical tool and will encourage him to use math in his own life.

No matter what curriculum you choose, remember to guard against independent thinking. Reinterpret every fact through a biblical, dependent philosophy, much like we did with the division lesson on page 42.

📖 *A Beka Arithmetic*—(www.abeka.com) A Beka offers a complete series of full-color textbooks written by Judy Howe. These textbooks predominately follow a drill approach to teaching math. They present math rules and facts, then drill the students on these rules through numerous problems. Due to this approach, these textbooks do not include much history or science.

📖 *Bob Jones*—(www.bjup.com) A good series to consider if you want a more textbook-based math curriculum. Although this series lacks the hands-on activities you would find in a more manipulative-oriented curriculum, the authors recognize that math is not neutral.

Curriculums and Resources 73

✋ ***Cornerstone Curriculum***—(www.cornerstonecurriculum.com) This K-Algebra curriculum strives to teach the student how to think mathematically and to understand what is really happening in math. It incorporates a fair amount of manipulative use as well as some textbook work. It does not, however, include much historical or scientific information. In the younger grades, the teacher's key tells the parent exactly what to say and do in each lesson. Some parents might find this detailed instruction helpful, but do not let it inhibit you from incorporating your own ideas into the curriculum.

✋ ***Horizons Math***—(www.aop.com) *Horizons Math* (produced by Alpha Omega Publications) seeks to combine analytical reasoning, hands-on learning, memorization, and drill. On some days the program incorporates lots of manipulatives into the program, while on other days it focuses on textbook drills and memorization techniques. The teacher presents the lesson, after which the student fills in a worksheet. While this series incorporates real-life manipulative use, it does not incorporate much history.

📄 ***Jacob's Math***—(www.whfreeman.com) Harold R. Jacobs has written an excellent textbook series for high school students. This series incorporates history and teaches the student to use math in problem-solving situations. This curriculum really stretches the thinking skills of the student, although it never connects math to God or the Bible.

📄✋***Key Curriculum Press***—(www.keypress.com) Key Curriculum Press offers several different sets of math curriculums that incorporate a great deal of pertinent information. This series does not purport to be Christian; however, they recognize that students need to learn math as something useful and, therefore, include a great deal of historical and practical information.

✋ Key Curriculum's ***Miquon Math*** is designed for elementary children. This curriculum has lots of potential, although it requires a good deal of parental involvement since the teacher's key lacks answers to all but the more challenging problems. Packed with useful scientific and historical information, this is an easy curriculum to adapt to a biblical perspective.

- **Key Curriculum Press** also offers a *Key to* series for late elementary and middle school students, which is laid out in an entirely different format, but maintains some of the same practical information as the Miquon Math. This series introduces everything in the student text and includes a teacher's key containing answers to every problem, keeping parental involvement minimal. Instead of coming in one large textbook, the *Key to* series comes in sets of smaller workbooks. Each set of workbooks covers a specific concept, such as fractions or measurements. This series was designed to supplement other curriculums, although some people may find that it also works well by itself.

- *Mastery Publications*—(www.masterypublications.com) Mastery Publications offers an affordable series for K through about 6th or 7th grade. One nice feature about this program is that parents are allowed to copy the workbook pages for several children. Another plus is that all the grades share one parent's manual that offers real-life activity ideas you can use to reinforce concepts. This program primarily focuses, however, on the memorization of math facts and skills, so you will need to be extra careful that you do not just present math as rote rules and skills.

- *Math-U-See*—(www.mathusee.com) This curriculum uses base 10 stacking block manipulatives and pictures to demonstrate the meaning of written numbers and operations. It comes with a video presentation of each lesson. The student builds *everything* he learns in math with manipulatives. Math-U-See is based on having the student build, write, and verbalize each math concept. It stresses place value and seeks to help the student understand how math works (instead of simply teaching him rules to memorize). While the program includes word problems, it does not contain a lot of real-life applications apart from the stacking block manipulatives. It also does not incorporate much historical information.

▣ *Mott Media*—(www.mottmedia.com) Mott Media has reproduced Ray's Arithmetic, a popular math series from the 1800s. The publishers also offer a Parent-Teacher's Guide by Ruth Beechick that shows today's parents/teachers how they can use the series. The guide divides the book into weekly lesson plans, offering ideas on games and ways to teach math outside of the book. The publisher recognizes that math is not neutral. Do not expect, however, to just hand your child the book and tell him to complete a certain number of pages. This program is designed to guide the *parent* in teaching math.

▣ *Saxon Math*—(www.saxonpublishers.com) The Saxon math series utilizes a review system to help students retain math concepts. Because of this review approach, the Saxon books jump around a lot between concepts. This layout, as well as other factors such as the lack of scientific and historical data, makes presenting each concept biblically particularly challenging. Saxon never claims to be Christian, so, as you might expect, these textbooks present math as an independent fact.

▣ *Simply Numbers*—(www.shoelacebooks.com) Shoelace Books sells a math course for grades K-5 called *Simply Numbers*. The curriculum consists of worksheets divided into daily lessons. Most lessons include practical word problems. Except on more complex concepts, the curriculum leaves concept presentation up to the parents. There is no official teacher's manual, but an answer key is provided.

♦♦♦♦♦♦

The list above by no means includes every math curriculum available. For additional curriculum thoughts, especially on high school curriculums, see James Nickel's recommendations in *Mathematics: Is God Silent?* and check out my website at www.christianperspective.net.

Biblical Reference Books

Mathematics: Is God Silent? by James D. Nickel (www.rosshousebooks.com) *Mathematics: Is God Silent?* is a valuable reference book for any math teacher. This extensive book functions like a Christian encyclopedia on math. While you may find this book challenging to read because of its thoroughness and mathematical terminology, it contains a lot of important mathematical information and is well worth the effort. In the first part of the book, you will find concepts and examples intertwined with a complete chronological history of math. The second part of the book explains how numerous math concepts reflect God's character and are used practically. The book also looks at teaching techniques and failures, offers guidelines for teachers, and evaluates a host of resources you can use to teach, especially if you have older children.

When you need more information about a math concept, look up the concept in the index and follow some of the links (try some from the beginning, middle, and end of the book). If the word you are looking for is not listed, try a similar or more general word. For example, when trying to find information about quotients, try looking up division. If you still cannot find what you are looking for, thumb through the book and see if a picture or heading looks related. Familiarizing yourself with this resource will help you find information about a concept quickly and easily.

Truth in the Transcendent by Larry L. Zimmerman (www.answersingenesis.org) This book offers a look at the thought and purpose behind math. It covers mathematical philosophies in technical and historical detail. By the end, the reader has no doubt that the Biblical approach to math is the only one that truly makes sense.

Supplemental Math Books

Below are a few supplemental math books that, while they do not necessarily claim to be Christian, contain some ideas you might find useful. Your library should also carry books similar to these that you can use. Please also check my website for a more current list.

Math For Every Kid: Easy Activities that Make Learning Fun by Janice Van Cleave
This book contains 101 math problems and experiments that combine math with science.

Math for the Very Young: A Handbook of Activities for Parents and Teachers by Lydia Polonsky
Games for Math: Playful Ways to Help Your Child by Peggy Kaye
Like the titles suggest, these books share simple activities to use with your younger math students (preschool to about third grade). Many of these activities are ones you do every day but might not have realized involve math. Some of the activities could also be adapted for use with older children.

The Only Math Book You'll Ever Need by Kogelman and Heller
This book explains how to balance checkbooks, figure out interest/loans, determine the yardage for a sewing project, and use math in other common settings. Although geared for high school to adult level, this book is comparatively easy to read and many of the concepts could be simplified for elementary/middle school students.

Real-World Math for Hands-On Fun! by Cindy A. Littlefield
This book provides ideas for younger children that integrate math with science, history, and real life.

Chapter 11:
Overcoming the Difficulties

As with anything worthwhile, teaching math will have its difficult moments. Sometimes you will be faced with personal obstacles or fears. Your child does not understand. You do not think you are good at teaching math. You are confused by the math concept yourself. You have run out of time or ideas. While teaching math may sometimes appear impossible, there are ways to handle the obstacles that arise.

Enormous words and phrases present the most common and largest obstacle to teaching math. You know the words I am talking about. The words you read several times without getting any closer to understanding them. The words that sound so intellectual that you ask yourself what you are doing reading the book in the first place. The words that make you check the title page to make sure you are not reading a Greek or Latin edition.

There have been times when I have read pages in math books that seemed impossible to grasp. I remember struggling for days trying to understand a particular concept in a math book. Once I finally understood it, I could hardly believe I had struggled so hard with such a simple concept.

I like to think of intellectual words as wrapping paper. Wrapping paper often hides useful gifts. Likewise, intellectual words often hide very simple, practical concepts. Remember that every part of math is a practical tool you can use to worship God and explore the earth. Each seemingly complex word you encounter disguises a very simple tool.

So do not be frustrated by words that are initially hard to comprehend. When you encounter a difficult word (or paragraph), stop and ask yourself what tool lies buried in that word.

Think back to God's simple truths about math. Think of how they relate to this specific area of math. Remember that the challenging word names some principle God created and sustains, or perhaps names a way mathematicians have agreed to write a problem. Reread the word with God's truths in mind, and you will be amazed at how simple even the most complex word becomes.

As you seek to explore math, you may encounter concepts that, try as you might, neither you nor your child can grasp. Do not panic! When you find yourself lacking knowledge, remember that your Heavenly Father is the source of all knowledge. "Wisdom and power are His" (Daniel 2:20). God knows all about math. Let Him teach you and your child.

I remember many times during my childhood that my mother and I could not understand an English concept. We would spend hours staring at it without success. Finally, Mom would bow her head and pray, telling God that we did not get it and asking Him to help me understand if I needed to know this concept. Often, after praying and looking at the lesson again, I would understand. Other times it would be made clear to me a few hours, days, or months later. One way or another, God *always* showed me what I needed to know.

If you do not think you have time to prepare biblical math lessons for your children, consider letting your children prepare the lessons with you. Walk through the errors in the lesson with them and let them help you find more information on how that particular concept should be approached biblically.

I remember my mom and I taking this approach to history lessons one year. No amount of prepared lessons could have taught me so much. Alongside my mom, I learned discernment and research skills. Above all, I watched her pray and ask God's help and strength when we both felt confused. Do not be afraid to learn and grow with your children.

Whether the difficulty you are facing bears the shape of an ominous word, a confusing concept, or a busy schedule, seek God's help and receive from Him the clarity, wisdom, and time you need. Make each new day of teaching exciting by relying on God for ideas and strength.

I still remember the fear I felt on one of my first skiing trips. As I slid off the ski lift and stared down the slope, my whole body started to tremble. I could never make it down this slope! The expert skiers around me called the slope easy, but it looked extremely difficult to me. I pointed my skis slightly downhill. The movement terrified me. I inched forward as slowly as possible. How could people actually enjoy skiing?

As I would later discover myself, people enjoy skiing because they stop focusing on themselves and their fears. They allow their eyes to take in the majestic mountain peaks around them. They breathe in the invigorating mountain air. Instead of letting the enormity of the slope before them overwhelm them, they tackle the mountain little by little. Although they may fall, they have learned to love the snow melting on their face and the challenge of standing up again. They actually relish the moguls that beginning skiers would term impossible obstacles.

Sometimes teaching math may feel as overwhelming as facing that steep slope. Although teaching math at first may seem impossible, do not panic and do not give up when you take a fall. Take your eyes off your fears. Instead, focus them on the Lord and on the breathtaking view of His character He has displayed in math. Rather than letting the enormity of your task overwhelm you, ask God to help you with the one math concept you need to teach at the moment. Just as I discovered the joys of skiing, you will discover God's joy and blessings in math. The very things about math that once intimidated you will eventually bring you great joy.

◆◆◆◆◆◆◆

> Always give yourselves fully to the work of the Lord, because you know that your labor in the Lord is not in vain.
>
> (1 Corinthians 15:58)

Teaching math biblically is worth all the effort! By showing your child how to approach math biblically, you are showing him how biblical truths affect everything he does. You are allowing him to experience the joy of seeing and using math as a practical tool. You are letting him explore God's creation and magnify his Creator. Above all, you are teaching him the *truth*. Instead of leaving your child with an independent understanding of math, you are letting math become a testimony to God in his heart. Do not be discouraged. Your labor in the Lord is *not* in vain.

Parting Thought

Exploring math with you has been a blessing. I have found myself challenged and encouraged by God's presence in math—challenged to trust Him more and encouraged that He will indeed be faithful to His Word. Yet I know that we have really only begun to see the ways math points us to God. Likewise, we have only touched on the teaching possibilities and opportunities that lie buried within math. My prayer for all of us is that we would leave here encouraged to further pursue God in math and that in math class, as well as in other areas of our life, we might—"Look to the LORD and His strength; seek His face *always*" (Psalm 105:4).

I would like to end by encouraging you to behold God in everything you do. Math is only one tiny area in which we can behold God. Every area of our life is, like math, a looking glass through which we can see God's reflection. The principles we have explored in math are equally true in science, history, English, and everything else we learn and do. God's truth is waiting for us *everywhere*. He longs for us to depend on Him in *everything*. May we fix our eyes on Him each day.

> But we all, with unveiled face, beholding as in a mirror the glory of the Lord, are being transformed into the same image from glory to glory, just as from the Lord, the Spirit.
> (2 Corinthians 3:18 NASB)

Appendix A: Idea Notebook

Listed below is a sample idea notebook containing a few common ways we use math in our daily lives. See chapter 9 to find out how to develop and use your own idea notebook as you teach.

Common Ways We Use Math

Addition/Subtraction:

- Balancing a checkbook
- Reconciling a bank statement
- Figuring out the age of something (how long a company has been in business, how long the luncheon meat has been in the refrigerator, how old someone is)
- Dealing with time problems (how long has dinner been in the oven, how long of a nap has Johnny had)
- Shopping

Algebra:

- Finding any sort of unknown value (Any time we find an unknown value using addition, subtraction, multiplication, or division we are really using algebra. For example, when we try to find the unit price of something, we are trying to find an unknown value. We could use a p, x, or any other letter to represent the unit price. Algebra simply uses a letter to represent the unknown in an equation.)
- Figuring out how to solve a problem (Sometimes we are not sure how to find the answer we need. For example, we might be trying to figure out how many

bags of mulch we need for the garden. We know that each bag covers 200 cubic feet and that we need to cover 1500 cubic feet. Thus we are trying to find what number times 200 will cover 1500 feet or $w \cdot 200 = 1500$. We take 1500 and divide it by 200 or $w = 1500/200$. Algebra is simply a different way of writing equations we do all the time. While most of us do not bother to write down the algebra behind everyday equations, algebra comes in handy when we do not know how to find the answer or when we are dealing with more complex equations.)

- Using a formula (finding the area of any shape, calculating interest rates)

Basic Number Functions:

- Telling time
- Reading radio dials, TV remotes, calculators, baking temperature gauges, thermometers, timers
- Dialing telephone numbers

Decimals:

- Dealing with money in any form
- Paying bills
- Balancing a checkbook
- Finding prices while shopping
- Dividing uneven amounts of any kind

See division for examples of things commonly divided.

Division:

- Dividing things evenly among family members (such as a bag of candy)
- Finding the price per unit (determining which size of spaghetti sauce is cheaper or determining the cost of each can of soda)

Appendix A: Idea Notebook

- Planning (figuring out how many pages in a book need to be read each day to finish it on time, finding how much yarn is needed for a crocheting/knitting project, deciding how much to lay aside each month for a yearly bill)

Fractions:

- Halving or doubling a recipe
- Cutting a pie or cake
- Converting between types of measurements (converting teaspoons to tablespoons or using a $1/4$ measuring cup instead of a $1/2$ measuring cup to measure $1/2$ cup of flour)
- Mixing solutions appropriately
- Measuring detergent or medicines
- Folding paper into halves or quarters for a card
- Determining yardage necessary for a sewing project

Geometry:
See measurement and graphing

Graphing:

- Reading a map
- Comparing any sort of results or data
- Planning a garden or other project

Measurement:
Home Improvements—

- Purchasing new flooring
  - Computing the area of a floor
  - Comparing prices between products that are sold differently

- Figuring out the best way to layout the material to avoid waste
- Converting square feet to square yards or vise versa
- Measuring windows for new drapes or blinds
- Preparing for new countertops/cabinets (how deep should the cabinets be, will the plumbing fit, what size sink is needed)
- Installing a light fixture (what size fixture will fit, is the fixture centered on the wall/ceiling)
- Finding the surface area of the walls in a room before painting it (to know how many gallons of paint to buy)
- Landscaping
  - Measuring the area to be landscaped
  - Purchasing the appropriate quantity of bark or rocks
  - Planning for plant growth and deciding how far apart to place plants

Around the House—

- Hanging a picture (especially if it needs to be centered)
- Arranging furniture (knowing if something will fit before it is moved)
- Cooking (uses many different measurements and conversions, such as teaspoons, tablespoons, cups, liters, ounces)
- Mixing solutions such as wallpaper stripper or miracle grow (can also involve multiplication if larger quantities are needed)

Multiplication:

- Determining the quantity of items necessary for a project (determining how many pencils are needed in order to provide five pencils per guest for 20 guests)

- Finding the total spent on something per year (finding how much is spent each year if $18 is spent on trash pick up each month)
- Shopping

Percents:

- Determining sales tax on an item
- Computing sale prices
- Comparing interest rates
- Grading papers (taking the total number of problems and figuring out the percentage of correct answers)

Rounding/Estimating:

- Shopping (keeping an approximate mental total)
- Estimating taxes/tax withholdings
- Budgeting
- Managing time (determining what time dinner will be ready if the crockpot started at 8:00 a.m., figuring out what time to leave for soccer practice in order to have time to stop at the bank on the way)

Sets/Counting:

- Organizing (putting like things together and sorting by sets)
- Doing the wash (sorting the dirty wash into sets by color or type, sorting the clean wash by family member, matching socks)
- Unloading the silverware (dividing into sets of forks, knives, big spoons, small spoons)
- Dealing with money (adding nickels, dimes, or quarters is really counting in sets of 5, 10, or 25)
- Counting a variety of objects (library books, people present, number of cookies, time, money, toys)

Common Projects That Use Multiple Math Concepts

- Sewing (uses fractions, multiplication, division, addition, subtraction, and even algebra)
- Woodworking (uses measurements, addition/subtraction, algebra, proportions, and graphs)
- Operating a lemonade stand (uses addition, decimals, division—adding the cost of ingredients and receipts uses addition; counting money uses decimals; finding the cost of each glass of lemonade involves division)
- Planning a party (uses multiplication and division—multiplication to determine how many items to purchase; division to find the cost per attendee)
- Buying a house or car (can involve almost every concept—finding and comparing interest rates involves percents and multiplication as well as decimal numbers; figuring out what numbers to divide/multiply uses an algebraic equation)

A Few Basic Ways Math Is Used in Science

- Classifying animals and plants
- Discovering order in creation (such as the order in sunflowers, honeycombs, red blood cells, and sound waves—see chapter 2)
- Comparing results of experiments (graph and compare the effects of different lighting on a seed's growth)
- Predicting reactions
- Converting measurements (liters to cups or milliliters)
- Mixing chemicals
- Designing and engineering products
- Predicting the results of a proposed design
- Selecting the best materials to use in a project

Appendix B: Bibliography

Modern Textbooks

The reviews given in chapter 10 were based on the textbooks listed below. Although other math textbooks were consulted, only reviewed textbooks are listed.

Beechick, Ruth. *Parent-Teacher Guide for Ray's New Arithmetics.* Milford, MI: Mott Media, 1985.

Cummins, Sareta A. *Horizons Mathematics 1 Teacher Handbook.* Chandler, AZ: Alpha Omega Publications, 1991.

Demme, Steven P. *Demme-stration Video.* Drumore, PA: Math-U-See, 1999.

———. *Math-U-See: Foundations of Mathematics.* Drumore, PA: Math-U-See, 1998.

———. *Math-U-See: Intermediate Mathematics Teacher's Manual.* Drumore, PA: Math-U-See, 1999.

Farmer, Letz. *Defeating Division.* Arden, NC: Mastery Publications, 1995.

———. *Finishing Fractions.* Arden, NC: Mastery Publications, 1991.

———. *Mastering Multiplication.* Arden, NC: Mastery Publications, 1990.

———. *Parent's Manual.* Arden, NC: Mastery Publications, 1997.

———. *Subduing Subtraction.* Arden, NC: Mastery Publications, 1990.

Hake, Stephen, and John Saxon. *Math 65: An Incremental Development.* 2nd ed. Norman, OK: Saxon Publishers, 1995.

———. *Math 76: An Incremental Development.* 2nd ed. Norman, OK: Saxon Publishers, 1992.

———. *Math 87: An Incremental Development.* Norman, OK: Saxon Publishers, 1991.

Horizons Mathematics K Teacher Handbook. Chandler, AZ: Alpha Omega Publications, 1994.

Howe, Judy England. *Arithmetic 1: Teacher's Edition.* With contributions by Patricia East, Kim Marie Ashbaugh, Cathryn Kauger, and Gloria Rigsby. Pensacola, FL: A Beka Book Publications, 1993.

———. *Arithmetic 3: Teacher Key.* 4th ed. Pensacola, FL: Pensacola Christian College, 1996.

———. *Arithmetic 4.* 1995 ed. Pensacola, FL: A Beka Book Publications, 1995.

———. *Pre-Algebra: Basic Mathematics II; With Problem Solving Strategies, Teacher Key.* 2nd ed. Pensacola, FL: A Beka Book Publications, 1996.

Jacob, Harold R. *Geometry.* 3rd ed. New York: W. H. Freeman & Co., 2003.

Jacobs, Tammie D., Susan J. Lehman, Dottie A. Oberholzer, Lynnette Chevalier, Lurene Dubois, Caryn M. Moody, Debra Overly, et al. *Math 2 for Christian Schools: Teacher's Edition.* 2nd ed. Greenville, SC: Bob Jones University Press, 1993.

Jacobs, Tammie D., Susan J. Lehman, Dottie A. Oberholzer, Lynnette Chevalier, Sharon Hambrick, Caryn M. Moody, Debra-Ann Overly, et al. *Math 3 for Christian Schools: Teacher's Edition.* 2nd ed. Greenville, SC: Bob Jones University Press, 1994.

———. *Math 4 for Christian Schools: Teacher's Edition.* 2nd ed. Greenville, SC: Bob Jones University Press, 1996.

Pilger, Kathy D., Ron Tagliapietra, Larry L. Hall, and Larry D. Lemons. *Algebra 1 for Christian Schools.* 2nd ed. Greenville, SC: Bob Jones University Press, 1999.

Quine, David. *Making Math Meaningful Level 1 Parent/Teacher Guide.* Rev. ed. Richardson, TX: Cornerstone Curriculum Project, 1997.

———. *Making Math Meaningful Level 6*. Rev. ed. Richardson, TX: Cornerstone Curriculum Project, 1998.

———. *My Very Own Making Math Meaningful Level K Parent Guide*. Rev. ed. Richardson, TX: Cornerstone Curriculum Project, 1998.

———. *Principles from Patterns: Algebra 1*. Rev. ed. Richardson, TX: Cornerstone Curriculum Project, 1997.

Rasmussen, Lore. *Miquon Math*. Emeryville, CA: Key Curriculum Press, 1978.

Rasmussen, Steve. *Key to Fractions*. Emeryville, CA: Key Curriculum Press, 1980.

Rasmussen, Steven, and David Rasmussen. *Key to Percents*. Berkeley, CA: Key Curriculum Press, 1988.

Rasmussen, Steven, and Spreck Rosekrans. *Key to Decimals*. Emeryville, CA: Key Curriculum Press, 1985.

Ray, Joseph. *Ray's New Practical Arithmetic*. Van Antwerp, Bragg & Co., 1877. Reprinted, Milford, MI: Mott Media, 1985.

Saxon, John H., Jr. *Algebra $^1/_2$: An Incremental Development*. 2nd ed. Norman, OK: Saxon Publishers, 1997.

———. *Algebra 1: An Incremental Development*. 2nd ed. Norman, OK: Saxon Publishers, 1990.

———. *Algebra 2: An Incremental Development*. 2nd ed. Norman, OK: Saxon Publishers, 1997.

Simply Numbers Grades 1-3. Stone Mountain, GA: Shoelace Books, 1999.

Older Textbooks

These textbooks dating from 1768-1930 were the basis for the generalities drawn in chapter 6.

Adams, Daniel. *Adams's New Arithmetic, in which the Principles of Operating by Numbers Are Analytically Explained....* Keene, NH: J & J.W. Prentiss, 1830.

———. *The Scholars Arithmetic; or, Federal Accountant. Containing . . . The Whole in A Form and Method Altogether New, for the Ease of the Master and the Greater Progress of the Scholar*. 5th ed. Leominster, MA: Adams & Wilder, 1804.

Austin, F. E. *Preliminary Mathematics*. Hanover, NH: 1917.

Barker, Eugene Henry. *Applied Mathematics for Junior High Schools and High Schools*. Boston: Allyn and Bacon, 1920.

Bonnycastle, John. *The Scholar's Guide to Arithmetic; or, A Complete Exercise Book for the Use of Schools with Notes, Containing the Reason of Every Rule. . . .* 9th ed. New York: Isaac Riley, 1815.

Breslich, Ernst R. *First-Year Mathematics for Secondary Schools*. 4th ed. Chicago: University of Chicago Press, 1915.

Brothers of the Christian Schools. *Primary Arithmetic Mental and Written*. New York: La Salle Bureau, 1900.

Fish, Daniel W. ed. *The Rudiments of Written Arithmetic: Containing Slate and Black-Board Exercises for Beginners, and Designed for Graded Schools*. New York: American Book Co., 1905.

Love, John. *Geodaesia; or, The Art of Surveying and Measuring Land Made Easy. . . .* Corrected and improved by Samuel Clark. London: 1768. A facsimile reprint of the 8th ed. Schenectady, NY: United States Historical Research Service, 1997.

Rich, A. W. *The New Practical Arithmetic: Presenting Definitions, Principles, Drills, Model Solutions and Test Problems*. Chicago: A. Flanacan Co., 1900.

Robinson, James. *The American Arithmetic: In which the Principles of Numbers Are Explained and Illustrated by a Great Variety of Practical Questions*. Boston: John P. Jewett & Co., 1850.

Root, Erastus. *An Introduction to Arithmetic: For the Use of Common Schools*. Revised, corrected, and enlarged. Norwhich: Russell Hubbard, 1811.

Seaver, Edwin P[liny]. *Logarithmic and Trigonometric Tables*. 1838.

Appendix B: Bibliography

Seaver, Edwin P[liny]., and George A. Walton. *The Franklin Elementary Algebra*. Boston: William War & Co., 1882.

Stone, John C., and James F. Millis. *A Secondary Arithmetic: Commercial and Industrial for High, Industrial, Commercial, Normal Schools, and Academies*. Boston: Benj. H. Sanborn & Co., 1908.

Yeingst, Wilbur M. *Practical Problems in Arithmetic: Book 1 for Third Grade*. St. Louis: Webster Publishing Co., 1930.

Young, J.R. *The Elements of Mechanics, Comprehending Statics and Dynamics. With A Copious Collection of Mechanical Problems. Intended for the Use of Mathematical Students in Schools and Universities. With Numerous Plates*. Revised and corrected by John D. Williams. Philadelphia: Hogan and Thompson, 1839.

Supplemental Math Books

Below are a some supplemental math books. See chapter 10 for details about a few of my favorites.

Adler, Irving. *The Giant Golden Book of Mathematics: Exploring the World of Numbers and Space*. New York: Golden Press, 1960.

Dalton, Leroy C. *Algebra in the Real World*. Palo Alto, CA: Dale Seymour Publications, 1983.

Easterday, Kenneth E., Loren L. Henry, and F. Morgan Simpson, comps. *Activities for Junior High School and Middle School Mathematics: Readings from the Arithmetic Teacher and the Mathematics Teacher*. Reston, VA: The National Council of Teachers of Mathematics, 1981.

Kaye, Peggy. *Games for Math: Playful Ways to Help Your Child*. New York: Pantheon Books, 1987.

Kogelman, Stanley, and Barbara R. Heller. *The Only Math Book You'll Ever Need*. New York: Facts on File Publications, 1994.

Littlefield, Cindy A. *Real-World Math for Hands-On Fun!* Charlotte, VT: Williamson Publishing, 2001.

Markle, Sandra. *Discovering Graph Secrets*. New York: Atheneum Books for Young Readers, 1997.

Polonsky, Lydia, Dorothy Freedman, Susan Lesher, and Kate Morrison. *Math for the Very Young: A Handbook of Activities for Parents and Teachers*. New York: John Wiley & Sons, 1995.

Riley, James, Marge Eberts, and Peggy Gisler. *Stand Up Math: 180 Fun and Challenging Problems for Kids!* Glenview, IL: Good Year Books, 1995.

Smoothey, Marion. *Let's Investigate Numbers*. New York: Marshall Cavendish Corporation, 1993.

———. *Number Patterns*. New York: Marshall Cavendish, 1993.

VanCleave, Janice. *Janice VanCleave's Math for Every Kid: Easy Activities that Make Learning Fun*. New York: John Wiley & Sons, 1991.

Vergara, William C. *Mathematics in Everyday Things*. New York: Harper and Brothers, 1959.

Miscellaneous

A handful of resources on everything from inventions to a biblical view of math.

Duffy, Cathy. *Christian Home Educators' Curriculum Manual: Elementary Grades*. Westminster, CA: Grove Publishing, 2000.

Educational Research Library (National Institute of Education). *Fifteenth to eighteenth century rare books on education: a catalog of the titles held by the Educational Research Library*. Washington: National Institute of Education: For sale by the Supt. Of Docs., U.S. Govt. Print. Off., 1976.

Gillen, Allen L. *Body by Design*. Green Forest, AZ: Master Books, 2001.

Hoskin, Michael, ed. *Cambridge Illustrated History: Astronomy*. Cambridge, UK: Cambridge University Press, 1997.

Appendix B: Bibliography

James, Peter, and Nick Thorpe. *Ancient Inventions*. New York: Ballantine Books, 1994.

Knott, Ron. "Fibonacci Numbers and Nature." University of Surrey. http://www.mcs.surrey.ac.uk/Personal/R.Knott/Fibonacci/fibnat.html

Mayall, R. Newton, and Margaret W. Mayall. *Sundials: Their Construction and Use*. 3rd ed. Cambridge, MA: Sky Publishing Corp., 1994.

Nickel, James D. *Mathematics: Is God Silent?* Rev. ed. Vallecito, CA: Ross House Books, 2001.

Newton, Isaac. *Universal Arithmetick; or, A Treatise of Arithmetical Composition and Resolution*. Translated by Mr. Ralphson and revised and corrected by Mr. Cunn. London: W. Johnston, 1769.

Reid, Struan. *Invention and Discovery*. Universal ed. London: Usborne Publishing Ltd., 1986.

Schlager, Neil, ed. *How Products Are Made: An Illustrated Guide to Product Manufacturing*. Vol. 1. Detroit: Gale Research Inc., 1994.

The Shorter Catechism. Agreed upon by the Reverend Assembly of Divines at Westminster. Quoted in *The New England Primer, Improved*. . . . Boston: printed by Edward Draper, Newbury Street and sold by John Boyle, Marlborough Street, 1777. Reprint, Aledo, TX: WallBuilder Press, 2003.

Zimmerman, Larry. *Truth and the Transcendent: The Origin, Nature, and Purpose of Mathematics*. KY: Answers in Genesis, 2000.

About the Author

Katherine Loop has authored various resources for Christians including several family devotionals. She has also spoken to moms on various topics. Having been homeschooled herself, Katherine has a special heart to encourage the homeschooling community to look to the Lord. She currently lives at home with her parents and brother in Northern Virginia where she teaches piano, writes, and oversees Christian Perspective.

About Christian Perspective

Christian Perspective's purpose is to encourage others through resources, free newsletters, email/phone support, speaking engagements, and any other method the Lord provides. You can find out more about us or about our products on our website, www.christianperspective.net. While you are there, drop us a line or email us at info@christianperspective.net.

Coming Soon...

The author of this book is curently working on a supplemental math curriculum parents can use to help them teach math biblically. When completed, this resource will present various math concepts from a biblical perspective. It's the author's goal to have this resource available by the spring/summer of 2007. Visit www.christianperspective.net for more information.

Katherine Loop   5/9/09

"Delighting in This Season:
    Encouragement for Young Ladies"

I. Lost in New   Ps. 55:22
    Decisions and
    Burdens

   She felt God had called her
   to stay at home under her
   father's authority.

   As she sought the Lord, He
   opened and closed the doors

   She realized she was accountable
   to Him.

   Ps. 139:16   "In Your book were
                  all written the days that were
                  ordained for me."

   Enjoy the "uphills" (roller coaster)

II. Wordly Thinking
      "I have come that they might
       have life." John 10:10

   • work — our definition of success is
     different from the world's.

   • getting it right — striving in our
        own strength to show our
        own righteousness.
     tree with real fruit vs. fake fruit.
   Col. 2:6 as you have received Him,
                so walk in Him.